Invertebrate Blood Cells

Invertebrate Blood Cells

Volume 1

General aspects, animals without true circulatory systems to cephalopods

Edited by

N. A. Ratcliffe and A. F. Rowley

Department of Zoology
University College of Swansea
Singleton Park
Swansea SA2 8PP
Wales

1981

ACADEMIC PRESS

A Subsidiary of Harcourt Brace Jovanovich, Publishers

London · New York · Toronto · Sydney · San Francisco

ACADEMIC PRESS INC. (LONDON) LTD.
24—28 Oval Road
London NW1 7DX

U.S. Edition published by
ACADEMIC PRESS INC.
111 Fifth Avenue
New York, New York 10003

British Library Cataloguing in Publication Data

Invertebrate blood cells.
  1. Invertebrates—physiology
  2. Blood cells
  I. Ratcliffe, N A   II. Rowley, A F
  592′.01′13   QL364   80-41248

ISBN 0-12-582101-8

LCCCN 80-41248

Filmset in 'Monophoto' Times New Roman by
Eta Services (Typesetters) Ltd, Beccles, Suffolk
Printed in Great Britain by
Galliard (Printers) Ltd, Great Yarmouth

List of contributors

ANDERSON, R. S.   *Sloan-Kettering Institute for Cancer Research, Donald S. Walker Laboratory, 145 Boston Post Road, Rye, New York 10580, U.S.A.*

BAUCHAU, A. G.   *Departement de Biologie Animale, Facultes Universitaires de Namur, Belgium.*

CHENG, T. C.   *Marine Biomedical Research Program and Department of Anatomy (Cell Biology), Medical University of South Carolina, Charleston, South Carolina 29412, U.S.A.*

COOPER, E. L.   *Department of Anatomy, School of Medicine, University of California, Los Angeles, California 90024, U.S.A.*

COWDEN, R. R.   *Department of Anatomy and Program in Biophysics, College of Medicine, East Tennessee State University, Johnson City, Tennessee 37601, U.S.A.*

CURTIS, S. K.   *Department of Anatomy and Program in Biophysics, College of Medicine, East Tennessee State University, Johnson City, Tennessee 37614, U.S.A.*

DALES, R. P.   *Department of Zoology, Bedford College, University of London, Regent's Park, London NW1 4NS, England.*

DIXON, L. R. J.   *Department of Zoology, Bedford College, University of London, Regent's Park, London NW1 4NS, England.*

DYBAS, L.   *Department of Biology, Knox College, Galesburg, Illinois 61401, U.S.A.*

FITZGERALD, S. W.   *Department of Zoology, University College of Swansea, Singleton Park, Swansea SA2 8PP, U.K.*

HAYWARD, P. J.   *Department of Zoology, University College of Swansea, Singleton Park, Swansea SA2 8PP, U.K.*

RATCLIFFE, N. A.   *Department of Zoology, University College of Swansea, Singleton Park, Swansea, SA2 8PP, U.K.*

RAVINDRANATH, M. N.   *Department of Zoology, University of Madras, Chepauk, Tamil Nadu 600 005, India.*

ROWLEY, A. F.   *Department of Zoology, University College of Swansea, Singleton Park, Swansea, SA2 8PP, U.K.*

SAWYER, R. T.   *Department of Molecular Biology, University of California, Berkeley, California 94720, U.S.A.*

SHERMAN, R. G.   *Department of Zoology, Miami University, Oxford, Ohio 45056, U.S.A.*

SMINIA, T.   *Vrije Universiteit, Biologisch Laboratorium, Amsterdam 1007mc, The Netherlands.*

SMITH, V. J. *University Marine Biological Station, Millport, Isle of Cumbrae, Scotland.*

STEIN, E. A. *Department of Anatomy, School of Medicine, University of California, Los Angeles, California 90024, U.S.A.*

VAN DE VYVER, G. *Université Libre De Bruxelles, Faculte des Sciences, Laboratoire de Biologie, Animale et Cellulaire, Ave. F. D. Roosevelt, 50, 1050 Bruxelles, Belgium.*

WRIGHT, R. K. *Department of Anatomy, School of Medicine, Center for the Health Services, University of California, Los Angeles, California 90024, U.S.A.*

Preface

Renewed interest in invertebrate "blood cells" has developed for a number of reasons. First, there has been a recent escalation in research into comparative immunology, including many studies on invertebrates. These animals provide relatively simple experimental models, they may supply clues to the ancestry of the lymphoid system, and they may have novel defence reactions not yet discovered in the more complex immune systems of vertebrates. Secondly, many invertebrates, but especially the molluscs and crustaceans, are now being extensively farmed to augment the food resources of man. Clearly, a better understanding of the host defence reactions of such species would help to avoid and overcome disastrous outbreaks of disease which are likely to occur under the artificial and potentially stressful conditions of commercial culture. Thirdly, many invertebrates act as vectors of parasitic organisms which are the scourge of mankind. The insects and molluscs, in particular, include species responsible for the transmission of malaria, sleeping sickness, filariasis, onchocerciasis and schistosomiasis. The means by which these parasites invade and multiply in their hosts and yet fail to elicit an effective "immune" response is now the subject of intensive research. Fourthly, with the increasing resistance of invertebrate pests to chemical pesticides, and the accumulation of these noxious substances at higher levels in the food chains, greater efforts are being made to develop and utilize biological control agents such as the viruses, bacteria, fungi, nematodes and parasitoids. The potentially immense practical value of such agents has provided yet a further stimulus for researches into the host defence reactions of invertebrates. Finally, many biologists are beginning to realize that invertebrate blood cells not only function in "immune" or host defence reactions but also serve, at least in some species, to store, transport and/or synthesize food, waste products and hormones and are thus involved in many other vital life processes. Our lack of knowledge of the role of invertebrate blood cells in these basic functions provides a major area for future research utilizing modern biochemical techniques.

   Much of the recent work on invertebrate blood cells, particularly in the fields of comparative immunology and cellular defence reactions, has been summarized in a number of excellent volumes including *Contemporary*

Topics in Immunobiology, Vol. 4, Invertebrate Immunology (E. L. Cooper, Ed., Plenum Press, 1974); *Invertebrate Immunity* (K. Maramorosch and R. E. Shope, Eds, Academic Press, 1975); *Comparative Immunology* (J. J. Marchalonis, Ed., Blackwell Scientific, 1976); *Comparative Immunobiology* by M. J. Manning and R. J. Turner (Blackie, 1976); and *Insect Hemocytes* (A. P. Gupta, Ed., Cambridge University Press, 1979). However, not since the publication of Warren Andrews *Comparative Hematology* (Grune and Stratton, 1965) has an attempt been made to summarize comprehensively our knowledge of the structure and function of invertebrate blood cells. Such a synopsis is urgently required in the light of current research interests and the many advances which have been made utilizing sophisticated modern techniques.

Volumes 1 and 2 of *Invertebrate Blood Cells* have been prepared specifically to bridge the gap since the publication of *Comparative Hematology*, with each chapter written by an expert in his or her particular group of animals. We have attempted to present information on as many invertebrate groups as possible and to this end have included much unpublished material, some of which was researched specifically for these books, e.g. Chapters 5, 14 and 15 on the leeches, lophophorates and echinoderms, respectively. We hope these volumes will generate further interest in many areas of invertebrate haematology and provide a source of comparative data for those workers already researching into particular aspects of invertebrate blood cells.

February, 1981   N. A. Ratcliffe and A. F. Rowley

Contents of Volume 1

Contents of Volume 2

Section IX: Comparative Aspects of the Structure and Function of Invertebrate and Vertebrate Leucocytes

Section I

Aspects of the Evolution and Development of Body Cavities, Circulatory Systems and "Blood Cells"

1. Aspects of the evolution and development of body cavities, circulatory systems and "blood cells"

R. P. DALES

Department of Zoology, Bedford College, University of London, Regent's Park, London NW1 4NS, England

CONTENTS

I. Introduction

While the invertebrates vary enormously in structure all must maintain the body in working order by tissue repair, by elimination of waste, and by internal defence against invading foreign organisms. Each phylum represents a basic body plan, a particular morphological theme with variations according to different life styles. But, whatever the body structure may be, all animals have cells capable of phagocytosis. Most of these cells are independent in the sense that they can move about, either by active migration or by passive transport in circulated body fluid. If phagocytes are to play a role in internal defence then they must be able to move to the right place when required. This presents no particular problem to an animal which is very small, but with increasing size there is an obvious advantage in having very many of these cells throughout the body or capable of being distributed when needed. The evolution of circulatory systems to convey both nutrients and to mediate respiratory exchange lent itself to this need, phagocytes or other defensive cells being distributed by the blood or body fluids. What we know of these cells, their types and functions, is summarized in the chapters which follow. This introductory chapter is designed simply to remind the reader of the morphology of the different invertebrates we shall be considering and, in particular, the relationships of the various body cavities and vascular systems in which the cells are found (Table I).

The phagocyte is an eating cell and antedates the evolution of body cavities and indeed organ systems. The universality and importance of the phagocyte was emphasized by Metchnikoff (1893) in his well known book "Lectures on the Comparative Pathology of Inflammation". The most primitive invertebrates are without discrete organ systems and in the siliceous and horny sponges, digestion occurs solely through the agency of phagocytes (Pourbaix, 1931). In sponges food particles are trapped by choanocytes, then passed to the amoebocytes which distribute the products of digestion throughout the body (Fig. 1). Sponges have no body cavity other than the central "spongocoel" and the canals leading to and from it. In coelenterates with a single internal cavity which functions as a gut, the epithelial lining has become a specialized digestive layer and the products of digestion are distributed through canals radiating from this central cavity. In larger coelenterates with extensive mesogloea formation there is a need for scavenging, and amoebocytes are found which act as phagocytes (see Chapter 2).

Choanocytes   Amoebocyte

Fig. 1. Amoebocyte taking up carbon grains directly from choanocytes in the calcareous sponge *Grantia compressa* (after Pourbaix, 1931).

The acoelous turbellarians have an amorphous "gut" without a defined epithelium. The wandering amoebocytes break down and distribute the products of digestion. It is but a simple step for such cells programmed to engulf non-self particles as food, to take on the role of eliminating unwanted materials, foreign organisms and parasites, in order to maintain fitness.

We are probably most familiar with the defence systems characteristic of advanced vertebrates, and in particular those of mammals and birds. We now understand a great deal about their elegant immune responses and the role of the different cell types found in the blood and lymph. Both mammals and birds are characterized by their high activity maintained by elevated, constant body temperatures with a high pressure blood vascular system connected with a low pressure lymphatic system draining the tissue fluid. Defence cells can thus be transported quickly via the blood to sites of injury and potential infection, while the lymph can participate in the distribution of other white cells concerned with the more long-term responses to foreign organisms or materials. Both the vascular and lymphatic systems develop within the embryonic mesoderm separately from the coelom which serves as a body cavity in which the viscera are slung. Unlike many invertebrates where the coelomic cavity is spaceous and in which the fluid may represent the major part of the animal's mass, the volume of the coelomic fluid in birds and mammals is very small as compared with that of the blood and lymph.

II. The coelom

In invertebrates the coelom developed as a space around the gut serving both to separate it from the muscular system and also to provide a hydrostatic skeletal system on which the muscles of such soft-bodied creatures could act. The coelom was thus a mechanical development which functioned primarily in relation to movement and locomotion. Subsequently the spaceous fluid-filled coelom assumed other functions including excretion and osmotic regulation. It also served as a space in which the gametes could develop.

III. The blood vascular system

The blood vascular system, on the other hand, developed in relation to two needs: first, to distribute products of digestion from the gut to other parts of the body, and secondly to convey the respiratory gases to and from the tissues. The adaptation of the circulation to these functions in invertebrates

TABLE I. Types of body cavities and vascular systems in different animal phyla.

Phylum	Common name	Chapter	Type of main body cavity	Type of vascular system
Porifera	sponges		absent	absent
Coelenterata	sea anemones, jelly fishes	2	absent	absent
Ctenophora	comb jellies		absent	closed
Platyhelminthes	flatworms		absent	closed
Nemertea	ribbon worms		absent	closed
Rotifera	rotifers		"pseudocoel" (= ? persistent blastocoel)	none
Nematoda	roundworms		"pseudocoel" (= ? persistent blastocoel)	none
Acanthocephala	thorny-headed worms		"pseudocoel" (= ? persistent blastocoel)	none
Endoprocta			"pseudocoel" (= ? persistent blastocoel)	none
Protostomes				
Annelida	earthworms, bristleworms	3, 4, 5	coelom	closed
Echiura	—	6	coelom	closed
Sipuncula	—	6	coelom	closed
Mollusca	slugs, snails, clams, octopus, etc.	7, 8, 9	haemocoel	open
Arthropoda	crabs, spiders, kingcrabs, millipedes, insects	10, 11, 12, 13	haemocoel	open
Phoronidea	—		coelom	closed
Bryozoa	—	14	coelom	absent
Brachiopoda	lamp shells		coelom	open

Deuterostomes				
Echinodermata	starfish, sea urchins	15		closed divided
Chaetognatha	arrow worms			absent
Pogonophora			coelom	closed
Hemichordata	acorn worms			closed
Chordata:				
1. Urochordata	sea squirts	16		open
2. Cephalochordata	lancelets	16		closed
3. Vertebrata	lampreys, hagfish, vertebrates			closed blood + lymphatic system

has been most ably reviewed by Martin and Johansen (1965). A vascular system became a necessity because of increasing body size while the coelom developed in relation to burrowing. Many of the earliest invertebrates must have lived in surface silts. The fluids contained in each system were moved about within the body, though for different reasons, and this provided a means of transport for the phagocytes. In an unspecialized and metamerically segmented invertebrate the blood vessels are conveniently arranged longitudinally with lateral vessels in each segment supplying the body wall. This simple plan may be modified in various ways in relation to particular hydrostatic needs, or the gametes may be restricted to certain segments. In a few annelids, the vascular system has been lost, the coelomic fluid and the coelomocytes then serving the needs of nutrient and gas transport. In the polychaete annelid, *Nephtys hombergii*, haemoglobin is dissolved in both blood and coelomic fluid, but the haemoglobins are different and this emphasizes the complete separation of the two systems (Weber, 1971).

IV. Origin of the coelom

The coelom is a cavity which lies in the mesoderm, and is lined with an epithelium or peritoneum from which the gametes often appear to originate. It was once argued because of this latter characteristic that the coelom was a "gonocoel" derived from enlarged gonadial cavities, the genital ducts representing coelomoducts. We now know, however, that in many invertebrates the genital cells are identifiable before the development of the coelomic epithelium with which they later become associated. The different ideas concerning the origin of the coelom have been discussed in detail by Clark (1964). The commonly accepted view now is that the association of the gonads with the coelom is secondary and that the coelom arose independently from that of metameric segmentation. Serial arrangement of body wall muscles derived from the mesoderm was a separate development in relation to locomotion. In coelomates, the coelom may arise either by separation of cells within the mesoderm (the schizocoel), or by evagination from the larval gut or archenteron (the enterocoel). The first method is known as schizocoelic, the second, enterocoelic (Fig. 2).

Schizocoelic cavities are recognizable even in some rhabdocoels. For Beklemishev (1969), blastocoels and schizocoels are primary body cavities without lining epithelia, while the coelom, however it develops, is a secondary cavity within its own epithelial lining. Others apply the word "schizocoel" to the coelom when it develops by cavitation within the mesenchyme. The annelid vascular system is thus primary, the coelom secondary and though both develop within the mesenchyme the annelid

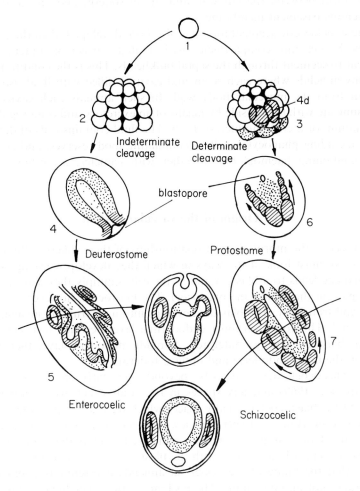

Fig. 2. Diagrams to show the origin of the coelom in typical protostomes (schizocoelic) on the right, and typical deuterostomes (enterocoelic) on the left. The fertilized ovum (1) cleaves to form a blastula by radial and horizontal cleavage (2) or by oblique (spiral) cleavage (3). In enterocoels, the mesoderm is derived from cells on each side of the blastopore (4). These dividing cells are carried inwards by growth on each side of the developing gut (archenteron) to form bands in which coelomic evaginations (5) develop and then separate as sacs surrounded by mesoderm (shown also in section). In protostomes, spiral cleavage gives rise to a pair of cells (4d) which become enclosed by ectoderm and endoderm. These 4d cells produce, by division, bands of mesoderm (6) the coelom arising as paired "splits" on each side of the developing gut (7) (shown also in section). Endoderm stippled; mesoderm hatched.

vessels form between the gut wall and the mesoderm and are lined by a collagenous basement membrane.

Where metameric segmentation has also been developed, then the coelom may be broken into compartments with limited transfer of contents from segment to segment through the septal bulkheads. This is the condition seen in many annelids where each segmental cavity is filled with fluid, isolating the gut from the muscular body wall. In such animals each cavity may communicate with the outside by means of an excretory and osmoregulating nephridium, or a coelomoduct, or both. The fluid contains cells of different types, including phagocytes, cells accumulating food reserves, and erythrocytes containing haemoglobin, together with gametes, according to the season.

V. Origin of the vascular system

To appreciate the morphological relationships of these different fluid-filled cavities, we must look at the ways in which they develop. Cleavage of the fertilized egg leads to a blastula and as the cells multiply they leave a fluid-filled space in the centre, the blastocoel. The blastocoel, however, is purely larval and transitory. The blastula develops into a gastrula by invagination at one end. This forms an inner endodermal layer destined to become the gut, the blastocoel being obliterated. A middle mesoderm layer then arises between the inner endoderm and outer ectoderm.

The blastocoel never persists beyond the early embryonic stage (Beklemishev, 1969) and any spaces in the adult which would appear to represent the blastocoel are really spaces which have developed later within the mesoderm. Hanson (1949) in reviewing different theories to account for the origin of the vascular system in annelids favoured the view that the vascular system is homologous with the blastocoel. The important point to note is that the lining of the vessels in annelids is essentially a basement membrane, not an epithelium. The coelom on the other hand is always a fluid filled sac or series of sacs surrounded by an epithelium or "peritoneum".

VI. Protostomes

We have already mentioned that there are two fundamentally different modes of coelom formation. In the protostomes cleavage is determinate, the mouth is formed at, or near the position of the blastopore, and the coelom develops as a schizocoel within bands of mesoderm derived from particular cells which may be identified during early cleavage (Fig. 2). The protostomes include all the lower invertebrates which have a trochophore larva or whose cleavage is basically spiral, such as annelids, the priapulids, sipunculids and echiuroid worms, as well as the molluscs. The lophophorate phyla (branchiopods, phoronids and bryozoans), and the largest invertebrate phylum, the arthropods, are also protostomes. Although the embryonic development of the insects is specialized it is derived from the determinate protostome type. The "pseudocoelomates" (rotifers, gastrotrichs, kinorhynchs, entoprocts, nematomorphs, acanthocephalans and nematodes) are also protostomes, but there is no general agreement about the homologies of body cavities in this heterogeneous assemblage of animals. The relationships of the pseudocoelomate phyla are far from clear, although the adults all have body cavities without an epithelium and, therefore, possibly a persistent blastocoel (Hyman, 1940), a view strongly opposed by Beklemishev (1969). We still know very little about the embryology of these invertebrates and grouping them together as "Aschelminthes" may indeed be artificial. It is obvious, however, that the need for body cavities is general and that the exact homologies from one group to another may not be clear since these structures develop in some way between ectoderm and endoderm. The pseudocoelomates are usually regarded as separate offshoots from the acoelomates. They are mostly small animals, and the body cavity must serve both for circulation and as an hydrostatic skeleton.

In annelids, the coelom is compartmented by the septa, but these do not prevent circulation of either fluid or cells (see Chapters 3, 4, 5). The cells arise from the lining epithelium probably from specialized tracts though very little is known about their origin. Further specializations in annelids are shown by the more advanced classes, the oligochaetes and leeches. In the oligochaetes, the genital function is commonly restricted to a few specialized segments and it is only in these that the coelom will communicate with the outside by means of coelomoducts. Nephridia are basically paired structures but again may be restricted, grouped together, or the nephridial ducts directed into the gut rather than through the body wall as an adaptation for water conservation. All these ducts have ciliated funnels which repel the cells so that these are not lost. Earthworms have, in addition, "dorsal pores" in each segment. It is through these that fluid and cells may be lost when the worm contracts violently.

In the leeches, the coelom is restricted by the development of packing mesenchymal tissue and forms a complex system of channels or sinuses. Since the blood vessels in annelids pass through the coelom, in some leeches the principal longitudinal vessels are enclosed within an outer (coelomic) sinus. The free cells in annelids are mostly, if not entirely, of coelomic origin so that cells in leeches will be found principally within these sinuses especially around the nephridia which connect the sinuses to the outside.

The coelom is further restricted in the molluscs but two functions of the coelom are retained. The first is to provide a genital space. The other function is to form a pericardial space around the heart. The coelom in molluscs is thus divided, one part being pericaridal and related to the kidney, the other being genital. The main body cavity is therefore not a coelom but a haemocoel, and the molluscan vascular system, unlike that of annelids, is usually open (see Chapters 7, 8). A haemocoel is a cavity which forms within the mesenchyme but it has no epithelial lining. The fluid within the haemocoel is known as "haemolymph". The heart acts as a circulatory pump and its muscles like those of the pericardial chambers are derived from the coelomic epithelium. The haemolymph contains dissolved haemo-proteins as well as cells which are therefore termed haemocytes. The coelomic fluid is restricted to the pericardial chamber and the gonads.

The differences between the body cavities in the annelids and molluscs probably resulted from the different locomotory mechanisms evolved. In both arthropods and molluscs the abandonment of a locomotory system combining the advantages of metameric segmentation and serial hydrostatic compartments led to a restriction of the coelom and development of an open blood vascular system and haemocoel. Although there are many detailed variations of this plan in the molluscs, as might be expected in such a diverse phylum, all molluscs possess a dorsal "heart" which maintains circulation of the haemocoelic fluid. In the cephalopods, the respiratory demands imposed by their activity have resulted in a circulatory scheme which approaches a closed system. The blood is confined to an extended vascular system with low-pressure sinuses together with vessels provided with accessory hearts which maintain the circulation to and from the gills. The other organs are surrounded by an extracellular fluid compartment with over twice the blood volume. This arrangement is remarkably like the blood and lymphatic system of vertebrates. Cephalopods also have distinct lymphatic "white organs" associated with the free cells (see Chapter 9).

The Onychophora, though arthropods, are of particular interest in showing many annelid features including metameric segmentation, coelomic pouches and nephridia (Chapter 10). These pouches are restricted and the heart resembles those of other arthropods, composed of a dorsal vessel with paired ostia in each segment and enclosed within a pericardial sinus. Contractions of the heart, pump the fluid forward, into the haemocoel.

In insects, the dorsal heart is the main propulsive organ, but secondary pumps sometimes occur near or within the appendages (see Chapter 13). The heart arises from cardioblast cells derived from two separate sites on the walls of coelomic sacs (Fig. 3). The cells come together and roll up to form the tubular heart. While the cardioblasts are, therefore, derived from the coelomic epithelium the lumen of the heart is not coelomic. The haemocytes

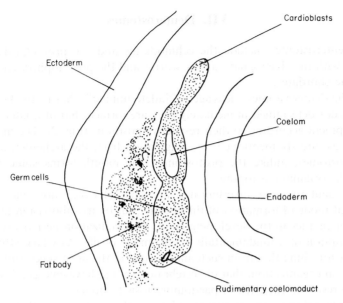

Fig. 3. Relationship of coelom to haemocoel in the developing stick insect *Carausius morosus* (after Cavallin, 1969).

all originate from the mesoderm and some apparently develop from cells originally part of the coelomic epithelium (see Chapter 13). The first to be formed arise from a thin median strip of mesoderm which lies above the developing nerve cord (Fig. 4). The insect haemolymph is thus, in origin, a mixture of coelomic and blastocoelic fluids and contains a variety of cells of different origin and function (Jones, 1977). They are termed "haemocytes" because they circulate within the haemocoel.

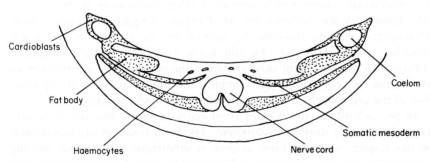

Fig. 4. Ventral part of a developing insect embryo. Diagram (based on Johannsen and Butt, 1941) showing relationship of the body cavities and origin of the haemocytes from the mesoderm.

VII. Deuterostomes

The deuterostomes include the echinoderms and the protochordates, together with the chaetognath arrow worms and the pogonophores as well as the true chordates.

In the deuterostomes, cleavage is "indeterminate", that is, the fate of the cells is not determined immediately they are formed, but at a later stage in development according to their respective positions (Fig. 2). The mesoderm forms, as already mentioned, by evagination from the archenteron. In the deuterostomes, unlike the protostomes, the mouth forms some distance from the original blastopore.

The deuterostome invertebrates are characterized, not only by the different mode of formation of the coelom, but by its subdivision into three pairs of cavities in the developing larva, a specialization perhaps related to the adoption of a fundamentally sessile mode of life. As Clark (1964) says "It is astonishing that from such animals one of the most successful types of structural organization, that of vertebrates, should have emerged". But they had to resort to the trick of paedomorphosis to do so.

In echinoderms, the tripartite origin of the coelom is reflected by the axocoel, the hydrocoel and the somatocoel. The echinoderm somatocoel forms the main body cavity supporting the viscera; the hydrocoel forms the water vascular system and is related to the ambulacral system, the left sac opening to the outside through the hydropore (Chapter 15). In asteroids the left axocoel and hydrocoel encloses the stone canal, and the ring vessel around the disk includes special organs known as Tiedemann's bodies from which the coelomocytes are believed to arise (see Chapter 15). The water vascular system is lined with epithelium and is ciliated internally to aid circulation of the fluid. In starfishes, the water vascular system constitutes the principal circulatory system, but there are, in addition, the haemal and the perihaemal systems which represent further subdivisions of the coelom. The haemal system includes the axial organ alongside the stone canal. Phagocytes circulate through the haemal system as well as the water vascular system. However, we still know very little about the precise functions of these different subdivisions of the coelom. The subdivisions of the coelom in echinoderms are complex, specialized and largely separate, but all the cells which circulate within them are coelomocytes.

In protochordates the coelom is similarly tripartite but this is mostly recognizable only during development. The hemichordates are exceptional in this respect, since the trunk coelom is subdivided in the adult and the coelomic spaces are commonly obliterated by parenchymatous tissue and muscles. In tunicates, the viscera lie in a true coelomic space with an epithelial lining. The pericardium is also coelomic and the heart is a fold in the pericardial wall. In hemichordates, the blood system consists of two longitudinal vessels rather reminiscent of the arrangement in annelids, but the blood is circulated by muscular contractions of the anterior pericardium, a typical deuterostome feature. In the cephalochordates, the vascular system is very like that of vertebrates in general arrangement, but the vessels lack an endothelium, and many are contractile. In spite of these differences few doubt that the vascular system of *Amphioxus* (*Branchiostoma*) is homologous with that of lower vertebrates.

VIII. Conclusions

The presence of an endothelial lining in chordates does not imply that the lumen of vertebrate vessels is coelomic, a paradox serving to emphasize the pitfalls awaiting the comparative morphologist in seeking to homologize different body cavities. We should, perhaps, be content when surveying the diversity of body cavities and vascular systems in the invertebrates, to recognize that all these structures must arise between ectoderm and endoderm, and that free cells wherever they are found are of mesodermal origin. While possession of phagocytes is probably universal, other cell-types will be related to the animals' needs, and distribution of the different types of cell within the body will depend on many factors. For example, the acquisition of an efficient circulation within a closed vascular system, with the development of a pump producing high blood pressures necessitated an efficient clotting mechanism (brought about by cell aggregation) or vaso-constriction (as in cephalopods) to prevent loss of blood when a vessel is damaged. On the other hand, prevention of fluid loss and microbial infection in an arthropod with a low-pressure haemolymph when the body wall is damaged, can be brought about by completely different mechanisms.

The terms applied both to body fluids and to the cells found within them are thus very variable and apparently similar cavities and cells may or may not be truly homologous. While "blood" and "lymph" may be reasonably distinct in higher vertebrates the term "blood" has often been applied somewhat indiscriminately to invertebrates, and "blood volumes" quoted in physiological literature often refer to a combination of blood and coelomic fluid. The haemolymph in arthropods may or may not contain dissolved haemoprotein, but the cells are often referred to as blood cells or "haemocytes". In annelids with a closed blood system one can call the cells in the blood "haemocytes" and the cells in the coelom "coelomocytes". Yet, the haemocytes may be coelomocytes which have migrated from the coelom into the blood vessels. If "haemocytes" are confined to a haemocoel then the cells in the blood of annelids are not haemocytes. Such etymological problems are unimportant providing the desire to find homologies is kept in perspective.

The various types of cell and the functions they perform in the major invertebrate groups are described in the chapters which follow. Elimination of waste, effete tissue, pathogen or parasite are often best effected by free cells. In view of the diversity of the invertebrates we should perhaps expect a diversity of solutions to achieve these ends.

References

Beklemishev, W. N. (1969). "Principles of Comparative Anatomy of Invertebrates". Vol. 2. Organology. Oliver and Boyd, Edinburgh.

Cavallin, M. (1969). C.r. hebd. Séanc. Acad. Sci. Paris **268**, 2189–2192.

Clark, R. B. (1964). "Dynamics in Metazoan Evolution". Oxford University Press, Oxford.

Hanson, J. (1949). Biol. Rev. **24**, 127–173.

Hyman, L. H. (1940). "The Invertebrates. Protozoa through Ctenophora". McGraw Hill, New York.

Johannsen, O. A. and Butt, F. H. (1941). "Embryology of Insects and Myriapods". McGraw Hill, New York.

Jones, J. C. (1977). "The Circulatory System of Insects". Charles C. Thomas, Springfield, Illinois.

Martin, A. W. and Johansen, K. (1965). "Adaptations of the Circulation in Invertebrate Animals". In: Handbook of Physiology (2) III Chap. 72, 2545–2581.

Metchnikoff, E. (1893). "Lectures on the Comparative Pathology of Inflammation". Kegan Paul, London.

Pourbaix, N. (1931). Notes Stat. Océanogr. Salammbo Tunis **23**, 3–19.

Weber, R. E. (1971). Netherlands J. Sea Res. **5**, 240–251.

Section II

Organisms Without Special Circulatory Systems

2. Organisms without special circulatory systems

G. VAN DE VYVER

Universite Libre De Bruxelles, Faculte des Sciences, Laboratoire de Biologie, Animale et Cellulaire, Ave. F. D. Roosevelt, 50, 1050 Bruxelles, Belgium

CONTENTS

I. Introduction

The life processes of an animal require that food and oxygen be continually available for metabolism and that wastes be promptly removed.

Complex animals, with organs and tissues far removed from the exterior or gut have a circulatory system to facilitate internal transport. Sponges, coelenterates, ctenophores, flatworms, however, have no circulatory system and simple diffusion serves to carry digested food, respiratory gases and waste products between various parts of their bodies. In nematodes, rotifers, entoprocts and gastrotrichs (= pseudocoelomates), a pseudocoelom exists in

19

which the body fluids are circulated and metabolic interchanges are aided by
streaming movements of the cytoplasm and by the endoplasmic reticulum
which provides a specialized pathway for intracellular transport.

In animals with no true circulatory system, wandering cells are present
and these have often been regarded as the progenitors of the coelomocytes
and blood cells (Andrew, 1965). However, despite some morphological
similarities, it is very difficult to establish a close parallel between the
functions of primitive wandering cells and of blood cells.

Thus, I shall not try to consider all the functions of wandering cell types
belonging to the primitive groups, but will rather attempt to point out the
means used by primitive groups to accomplish functions normally realized
by coelomic and blood vascular systems.

These functions include: (1) respiratory gas exchanges; (2) water and
digested food transport; (3) food storage; (4) elimination of organic wastes;
(5) "hormone" transport; (6) immunological defence.

II. Porifera

Sponges are filter-feeding animals which utilize flagellated cells to pump a
water current through their bodies. The sponge body is composed of an
external perforated epithelium, the pinacoderm, aquiferous channels and
choanocyte-chambers which move the water. Between these two layers is a
third region, the mesohyle, which includes an intercellular matrix, wandering cells and skeletal material (Fig. 1).

The consequence of such a structure is that there are not generally great
distances between the external and internal surfaces and the deeper-lying
cells or other elements of the body. Respiratory exchange between the water
current and the cells, where oxygen is utilized, takes place by diffusion. In
sponges there is a large area of naked cell surface available for exchange
along all the inhalant channels over the general body surface and at the
collar tentacles of the choanocytes.

The constant motion of water in sponges is maintained by beating
choanocyte flagella. This stream of water is not only nutritive but also
brings a supply of oxygen and removes CO_2 and other metabolic waste
products. Consequently, this flow of water assumes, depending on the
species, two main functions of the circulatory system, namely, gaseous
exchange and nitrogenous waste elimination.

Most of the wandering cells characteristic of sponges have a phagocytic
activity. One particular cell type, namely the archaeocyte, appears to be
very rich in phagosomes and secondary lysosomes (Fig. 2). These cells
receive nutritive material from the choanocytes and digest it. However,

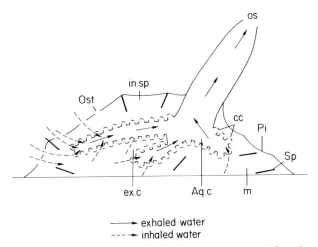

Fig. 1. Schematic section of a young fresh-water sponge (after Rasmont, unpublished). Ost, ostium; in.sp, inhalant space; ex.c, exhalant canal; os, osculum; Pi, pinacoderm; Sp, spicule; Aq.c, aquiferous channel; m, mesohyle; c.c, choanocyte chambers.

whether the archaeocytes carry the digested products directly to other cells or whether it diffuses through the mesohyle matrix is poorly understood. Indeed, the only accurate information is given by the study of Schmidt (1970) on *Ephydatia fluviatilis* who monitored the passage of fluorescent material from the archaeocytes towards the mesohyle 1 h after feeding.

In the animal kingdom, glycogen is generally the energy source used during the metabolic and morphogenetic processes. In the vertebrates, glycogen storage takes place in the liver hepatocytes. The liver branches off the circulatory system and may either discharge into the blood the glucose necessary for the metabolic processes or, alternatively, store the excess.

In sponges, there is no special organ for glycogen storage nor a circulatory system, nevertheless, one particular cell type, the glycocyte, is specialized for synthesis, storage and transfer of glycogen to regions of high demand; e.g. areas of growth, budding etc. (Boury-Esnault, 1977; Boury-Esnault and Doumenc, 1979). The glycocytes are characterized by an active Golgi, the presence of osmiophilic inclusions, α-glycogen rosettes, endoplasmic reticulum and mitochondria (Figs 3, 4). The morphology and the glycogen content of the glycocytes depends directly on the physiological state of the sponge tissue. In areas involved in morphogenetic processes, most of the glycocytes are phagocytized by archeocytes. In these cases, they appear to be a preferential nutritive source of archeocytes.

In metazoa, distant intercellular communication is generally brought

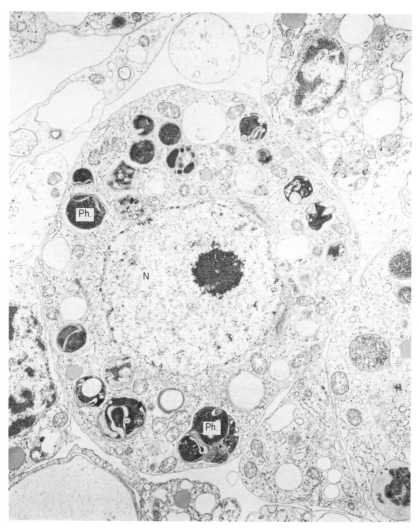

Fig. 2. Archaeocyte of the fresh-water sponge *Ephydatia fluviatilis* showing the presence of several phagosomes (N, nucleus; Ph, phagosome). × 8250. (Courtesy of Dr. M. Buscema).

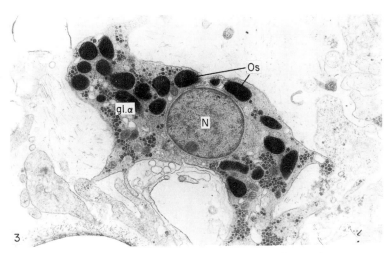

Fig. 3. Glycocyte of the marine sponge *Polymastia mamillaris* (gl. α, α-rosettes of glycogen; Os, osmiophilic inclusions). ×11 000. (Courtesy of Dr. N. Boury-Esnault).

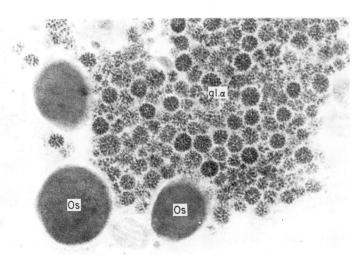

Fig. 4. Higher magnification of a *Polymastia mamillaris* glycocyte. Showing many α-glycogen rosettes (gl. α) and osmiophilic inclusions (Os.). ×45 000. (Courtesy of Dr. N. Boury-Esnault).

about by chemical messengers, e.g. hormones, which move to the target cells via the circulatory system. In sponges, these functions appear to be carried out by the mesohyle which constitutes a medium within which the diffusion of hormone-like substances is controlled. The role of such diffusible substances has been poorly studied, except in some special cases, e.g. the hatching and formation of gemmules in fresh-water sponges, or cell aggregation processes.

Gemmules are aggregates of several hundred cells included in a protein-aceous shell and which carry out asexual reproduction. It is known that the gemmules of some species are unable to hatch, as long as they remain included in the living mother sponge (Rasmont, 1975). This inability depends on the presence of an inhibitory factor, referred to as "gem-mulostasin". The chemical nature of gemmulostasin is unknown, but preliminary experiments indicate that it is a thermolabile, dialysable, pronase-resistant molecule, the molecular weight of which is less than 1000 daltons. Simultaneously, several experiments and observations strongly suggest that the aggregation of the gemmule cells is under the control of another diffusible substance (Rasmont, 1974), but such a substance has never been isolated.

The remarkable ability of sponges to aggregate after mechanical dissoci-ation was discovered early this century by Wilson (1907). Recent experi-mental work shows that such aggregation phenomena are brought about by an aggregation factor released, probably from the intercellular matrix, by sponges during their dissociation (Humphreys, 1963; Moscona, 1963; Van de Vyver, 1975). This factor appears to be a complex 70 S glycoprotein but it is yet to be discovered how it joins with the cells and with what degree of specificity. Moreover, this factor has not so far been shown to be involved in any recognition process that takes place in the normal life of the sponge. Meanwhile, the main features of coalescence or non-coalescence between sponges belonging to the same or to other strains can be explained, following the hypothesis of Curtis and Van de Vyver (1971), in terms of aggregation factors. These authors, demonstrated that the promoting factor of one strain increases the adhesiveness of homologous cells and decreases the adhesiveness of cells belonging to other strains. This would mean that the aggregation factors would play a role in the defence of sponges by protecting their integrity.

From this review, it can be concluded that in sponges, the absence of a circulatory system is palliated by three elements: (a) the water current, which is involved in both nutrient and O_2 supply, and the elimination of CO_2 and other waste products; (b) the wandering cells, which are mainly concerned with food transferance or storage; (c) the intercellular matrix, which appears to be the centre for long distance communications.

III. Coelenterata

The coelenterates are at the tissue grade of construction. They are tentacle-bearing metazoa, composed of a biepithelial body wall enclosing a central gastrovascular cavity (Fig. 5). This cavity has a single opening and, in contrast to sponges, no water current flows through the body. Nevertheless, when the mouth opens, water is drawn into the gastrovascular cavity, bathing the internal surface so that the cavity combines both circulatory and digestive functions.

A circulatory system is unnecessary in small animals like *Hydra*, where no cells are far removed from the external medium or from the gastrovascular cavity. In such cases, direct diffusion between the body and the surrounding water is sufficient to provide oxygen and to remove carbon dioxide and

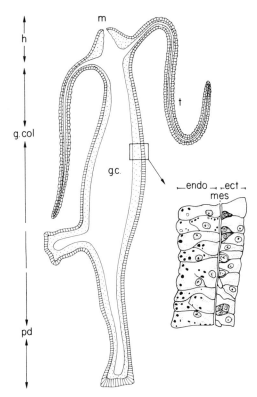

Fig. 5. Schematic section of a *Hydra*. g.c., gastric cavity; t, tentacle; m, mouth; h, hypostome; g.col, gastric column; pd, peduncle; ect, ectoderm; mes, mesogloea; endo, endoderm.

nitrogenous wastes. Some sort of transport system, however, is required in large solitary individuals or in colonies with an extensive mesogloea. The size limits set by diffusion of oxygen and carbon dioxide can be greatly extended if the fluids are in motion. This can be generated by body movements and by the gastrodermal flagella, which move the gastrovascular fluid through the cavity and its branches. Few muscular adaptations to move the fluid are found, but *Sertularia* polyps may have a muscular gastric pouch which contracts to force food into the coenosarc, and *Tubularia* polyps use the basal part of the hydranths to pump the food into the stem and roots.

Transport is still a more severe problem in the large medusae where the gastrovascular cavity is canalized so that the propulsion of the coelenteric fluid by the flagella is considerably increased. In some species, the canal systems tend to become truly circulatory in function, particularly in species with digestion restricted to a gastric cavity. In jelly-fish, both hydrozoan and scyphozoan, hypertrophy of the mesogloea for locomotion has transformed the gastric cavity from a single central space into a series of narrow canals.

The most effective circulatory devices are found in the Anthozoa where siphonoglyphs and septal flagella circulate the coelenteric fluid. That fluid circulation is a critical problem is shown by the differentiation of zooids as special circulatory individuals. These siphonozooids produce strong currents, sufficient to keep fluid moving in a complex system of solenial tubes.

The fluid that circulates in the gastrovascular system of coelenterates is unlike blood but it does contain dissolved or particulate food, mucus, enzymes, wastes, e.g. ammonia, dissolved gases and possibly coordinating or hormone-like substances.

Free wandering cells are poorly represented in coelenterates. They are completely absent in the Hydrozoa which are characterized by a very thin mesogloea. In many medusae, there are no wandering cells and when they exist, their qualification as "blood cells" remains uncertain. In particular, interstitial cells (Fig. 6), which are a self-proliferating line of undifferentiated RNA rich cells and which are primarily located between the epithelio-muscular cells of the ectoderm, can hardly be regarded as blood cells since their main function is to replace worn-out cells such as cnidocytes (Brien and Reniers-Decoen, 1949; Lehn, 1951), gland cells and most probably other cell types.

Glycocytes, very similar to those described for sponges, have been reported only in actinarians. They consist of cells wandering in the mesogloea (Van Praet and Doumenc, 1975; Doumenc, 1976), which contain variable amounts of α- and β-glycogen (Fig. 7).

As suggested by Chapman (1966) the mesogloea, despite its high water

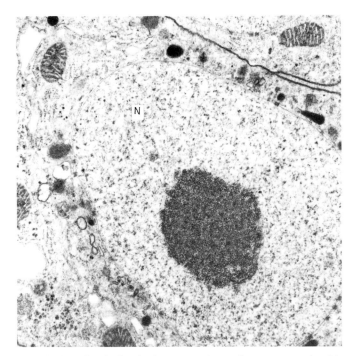

Fig. 6. Interstitial cell of the hydrozoan, *Craspedacusta sowerbi*, N, nucleus. × 32 250. (Courtesy of Dr. C. Massin).

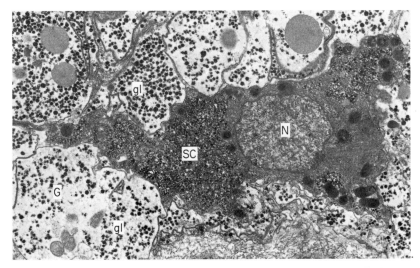

Fig. 7. Glycocyte of the actinian, *Cereus pedonculatus*. (G, glycocyte; gl, β-glycogen; SC, spermatic cyst; N, nucleus). × 7000. (Courtesy of Dr. D. Doumenc).

content (in some jelly-fish, it rises to about 99%), can hardly qualify as a circulatory system but it can be regarded as an over-developed intercellular material through which metabolites must be able to pass quite freely. It is worth noting that in the great majority of the coelenterates, the ionic composition of the mesogloea, the coelenteric fluid and the external medium are very similar (Table I) and that the degree of ionic regulation is very low.

TABLE I. Comparison of the ionic content of sea water and the mesogloea of the jelly-fish, *Aurelia* (after Robertson, 1949).

	Na^+	K^+	Ca^{2+}	Mg^{2+}	Cl^-	SO_4^{2-}	Protein
Sea water	478·3	10·13	10·48	54·5	558·4	28·77	—
Aurelia mesogloea	474·0	10·72	10·03	53·0	580·0	15·77	0·7

The existence of chemical messengers has been demonstrated in recent years, particularly in the case of *Hydra* (Schaller, 1973; Schmidt and Schaller, 1976; Berking, 1977), but nothing is known about the origin and the means of transport of these "hormone like" substances.

Finally, as in sponges, it does not seem that any specialized cell type is particularly involved in recognition and rejection phenomena which are known to occur in coelenterates at both allogeneic and xenogeneic levels (Theodor, 1970; Hildemann *et al.*, 1977). However, recently Patterson and Landolt (1979) have shown that injury responses in the anthozoan, *Anthopleura elegantissima*, the immigration of a specific cell type possibly occurs into the wound where it facilitates repair by laying down protein-aceous materials. This may indicate the presence of some chemical control system which pre-dates the origin of a circulatory system (Patterson and Landolt, 1979).

IV. Platyhelminthes

Platyhelminthes differ greatly from sponges and coelenterates since they possess a mesodermal layer to which "higher" animals have delegated an important share of organogenesis, e.g. formation of circulatory system etc. Nevertheless, Platyhelminthes are still devoid of organs which specifically accomplish this function and, in consequence, flat-worms suffer severe physiological problems. The mesodermal layer contains many more cells than the mesogloea and has higher metabolic requirements. In *Planaria*, for example, the oxygen consumption is ten times higher than in a coelenterate on a g/h basis, yet they are scarcely better equipped than coelenterates to provide oxygen or eliminate carbon dioxide. In most cases, however, the body surface is large enough to allow for all the respiratory gas exchanges necessary. Moreover, in some species, e.g. planarians, the supply of oxygen is assisted by the movement of cilia on the body surface (Vernberg, 1967).

Generally, most of the Platyhelminthes have resolved their internal transport problems by developing an exceptionally large surface/volume ratio for most of their organs which are bathed in the parenchymal tissue fluid. For instance, the gastrovascular cavity is highly branched, ramifying into all parts of the body. It contains a fluid that is circulated in a rather irregular fashion by the contraction of muscles in the body wall surrounding the cavity. Such internal circulation certainly aids respiratory gas transport and food distribution, but it is much less efficient than an organized circulatory system in which blood is pumped through definite channels.

In Platyhelminthes the presence of haemoglobin, which is a tissue pigment, is an exception rather than a rule. It has been reported for only two species of turbellarians and eight species of trematodes (Vernberg, 1967). Nothing is known of the origin of trematode haemoglobins and their function is unclear. Manwell (1960) has stated that although oxygen storage is the basic function of tissue haemoglobins, nevertheless, the amount of oxygen stored in the haemoglobin would not allow the organisms to maintain aerobic metabolism for any length of time under anoxic conditions. It is worth recalling that for a relatively large number of parasites, e.g. among Trematoda and Cestoda, in the adult stage, there is no oxygen requirement for the normal physiological production of energy.

In the Platyhelminthes, the elimination of nitrogenous wastes needs specialized organs which in this group consists of a protonephridial or flame cell system. These lack any connection with a circulatory system, are probably inefficient, and indeed little is known about their importance in excretion (Read, 1968). However, the fact that soft-bodied animals such as flatworms are able to live in brackish and fresh-water environments indicates that they must exercise some control over the osmotic uptake of water. The protonephridial system is poorly-developed in marine species and highly-developed in fresh-water forms which seems to indicate that the main function of these primitive excretory organs is osmoregulation.

The mesodermal parenchyma, which surrounds the internal organs, is composed of large vacuolated cells, collagen fibres, muscles and some wandering cells. These amoeboid cells, the so-called neoblasts, have often been compared to the coelenterate interstitial cells (Fig. 6) because of their totipotential capacities. But like them, despite the fact that their ultrastructure resembles that of the vertebrate haemocytoblasts, they are not comparable to blood cells (Andrew, 1965).

V. Pseudocoelomates

Nematodes are with the rotifers, the entoprocts and the gastrotrichs some of the most primitive animals which possess a general body cavity, but they are still devoid of a true vascular system transporting blood. In these groups, the general body cavity is a pseudocoelom, that is, a space formed between the body wall and the gut. It is fluid-filled and contains a number of mesenchymal cells. In adult worms, most of the cells form a network and a few, called pseudocoelomic cells, also remain free. The pseudocoelomic cells are not phagocytic and their function is unknown.

Because of the lack of any circulatory system the pseudocoelom plays an important role in internal transport. For example, food digested in the gut and absorbed in the intestine must diffuse through the pseudocoelomic fluid to reach all the cells, just like the respiratory gases which pass through the body wall. No nematodes possess differentiated respiratory organs. The supply of oxygen offers little problems in many small free living worms. In some habitats, however, like anaerobic muds or the digestive tracts of other animals, the supply of oxygen is inadequate to provide for a fully aerobic metabolism. In such cases, either anaerobic metabolic processes can occur, or respiratory pigment may be used to capture what little oxygen is available. The high affinity of nematode haemoglobins for oxygen suggests that they may function as oxygen carriers. For instance, in *Ascaris*, a large parasitic species, there are two different haemoglobins: one in the muscles of the body wall and the other in the pseudocoelomic fluid (Davenport, 1949; Smith and Morrison, 1963). The haemoglobin of the body wall only acts as an oxygen carrier at very low pressure, such as might occur in a vertebrate gut. In contrast, the pseudocoelomic haemoglobin does not seem to be involved in respiration since the oxygenated form does not dissociate under ordinary physiological conditions (Rogers, 1969). Nothing is known about the synthesis of haemoglobin in nematodes, but it has been proved experimentally that *Ascaris* requires an external source of haematin in order to maintain its haemoglobin content (Smith and Lee, 1963). Nevertheless, it can be said that in *Ascaris* and in many parasitic forms, there is no oxygen requirement for the normal physiological production of energy in the adult stage.

Nematodes are predominantly ammonotelic, particularly species which live in an aqueous medium. But Cavier and Savel (1954), however, found that when access to water over the body surface of *Ascaris* was restricted, the proportions of nitrogenous compounds in the excreta were changed. The ammonia nitrogen fell from 70 to 27% and the urea nitrogen rose from 7 to 51%.

Nematode excretory systems are composed of a single gland-cell with a duct leading to the excretory pore and to a network of lateral blind tubules. Despite the fact that these systems are referred to as excretory canals, an excretory function has never been firmly established. In adult nematodes, only in a few cases has evidence been presented for excretion of metabolic wastes by this system. Indeed, it seems that, generally, nitrogenous wastes are released into the media from the anus and not from the excretory pore. Moreover, in some species, the so-called excretory system is totally absent (Rogers, 1969).

Similarly, osmotic and ionic regulation does not seem to be related to the excretory system or the pseudocoelomic fluid but is the function of the digestive tract.

VI. Concluding remarks

To conclude, we can say that despite their low level of organization, primitive groups, which lack any form of circulatory system, have compensatory adaptations to assist internal transport and free them from complete dependence upon diffusion and intracellular transport. Free wandering cells, which are often associated with circulatory systems, are in fact poorly represented in the non-coelomic groups, and when they exist, they can hardly be regarded as true blood cells since their function is generally specific to their own group.

References

Andrew, W. (1965). "Comparative Hematology". Grune and Stratton, New York and London.
Berking, S. (1977). *Wilhelm Roux Arch. EntwMech. Org.* **181**, 215–225.
Boury-Esnault, N. (1977). *Cell Tiss. Res.* **175**, 523–539.
Boury-Esnault, N. and Doumenc, D. (1979). *In* "Biologie des Spongiaires". Colloque international CNRS 291.
Brien, P. and Reniers-Decoen, M. (1949). *Bull. biol. Fr. Belg.* **51**, 33–110.
Cavier, R. and Savel, J. (1954). *Bull. Soc. Chim. biol.* **36**, 1425–1432.
Chapman, G. (1966). *In* "The Cnidaria and Their Evolution" (W. J. Ress, Ed.) *Symp. Zool. Soc. Lond.* **16**, 147–168.
Curtis, A. S. G. and Van de Vyver, G. (1971). *J. Embryol. exp. Morph.* **26**, 295–312.
Davenport, H. E. (1949). *Proc. R. Soc.* B **136**, 225.
Doumenc, D. (1976). *Arch. Zool. exp. gén.* **117**, 295–324.
Hildemann, W. H., Raison, R. L., Cheung, G., Hull, C. T., Alaka, L. and Ikamoto, J. (1977). *Nature* **270**, 219–223.
Humphreys, T. (1963). *Devl. Biol.* **8**, 27–47.
Lehn, H. (1951). *Z. Naturf.* **66**, 388–391.

Manwell, C. (1960). *A. Rev. Physiol.* **22**, 191–244.

Moscona, A. A. (1963). *Proc. natn. Acad. Sci., U.S.A.* **49**, 742–747.

Patterson, M. J. and Landolt, M. L. (1979). *J. Invertebr. Pathol.* **33**, 189–196.

Rasmont, R. (1974). *Experientia* **30**, 792–794.

Rasmont, R. (1975). *In* "Current Topics of Developmental Biology" (A. A. Moscona and A. Monroy, Eds), Vol. 10, pp. 141–159. Academic Press, New York and London.

Read, C. P. (1968). *In* "Chemical Zoology" (M. Florkin and B. T. Scheer, Eds), Vol. 2, pp. 328–357. Academic Press, New York and London.

Robertson, D. (1949). *J. exp. Biol.* **26**, 182–200.

Rogers, W. P. (1969). *In* "Chemical Zoology" (M. Florkin and B. T. Scheer, Eds), Vol. 3, pp. 379–426. Academic Press, New York and London.

Schaller, H. C. (1973). *J. Embryol, exp. Morph.* **29**, 27–38.

Schmidt, I. (1970). *Z. vergl. Physiol.* **66**, 398–420.

Schmidt, T. and Schaller, H. C. (1976). *Cell diff.* **5**, 151–159.

Smith, M. H. and Lee, D. L. (1963). *Proc. R. Soc. B.* **157**, 234–242.

Smith, M. H. and Morrison, M. (1963). *Biochim. biophys. Acta* **71**, 364–369.

Theodor, J. L. (1970). *Nature, Lond.* **227**, 690–692.

Van de Vyver, G. (1975). *In* "Current Topics of Developmental Biology" (A. A. Moscona and A. Monroy, Eds), Vol. 10, pp. 123–140. Academic Press, New York and London.

Van Praet, M. and Doumenc, D. (1975). *J. Micr. et Biol. cell.* **23**, 29–38.

Vernberg, W. B. (1967). *In* "Chemical Zoology" (M. Florkin and B. T. Scheer, Eds), Vol. 3, pp. 359–393. Academic Press, New York and London.

Wilson, H. V. (1907). *J. exp. Zool.* **5**, 245–258.

Section III

Annelids and Related Phyla

3. Polychaetes

R. P. DALES AND L. R. J. DIXON

Department of Zoology, Bedford College, University of London, Regent's Park, London NW1 4NS, England

CONTENTS

I. Introduction

In common with other annelids, the polychaetes have a true coelom. In many species it is capacious and houses a population of free cells which participate in the transport and storage of oxygen and metabolites, in excretion, and in defence against invasion by foreign organisms; functions

35

more usually associated with the blood and blood cells of higher animals. The coelom is also the site of gamete maturation, and the coelomic free cells or "coelomocytes" are in many cases involved in supplying nutrients to the developing germ cells.

Most polychaetes are provided with a closed blood system but this may be reduced or entirely absent. The vascular respiratory pigment, if present, is usually in solution, and blood cells or "haemocytes" have only been recorded in relatively few species. In consequence, most of the research on polychaete free-cells has been upon the coelomocytes although some authors refer to them as "blood" cells.

The first detailed description of polychaete coelomocytes was published by Williams in 1852. He examined a number of species including *Phyllodoce lamelligera*, *Nephtys hombergii* and *Glycera alba* and observed that their coelomic fluid was more or less corpuscular while he detected no cells in the blood. Other early observations were made by Quatrefages (1865) and Claparède (1861, 1864) who drew attention to the red cells which occur in the colourless coelomic fluid of capitellids. Much of the early work was reviewed by Kollmann (1908) and Romieu (1923).

Polychaetes are morphologically and physiologically an extremely varied group of animals and their free cells exhibit a diversity of structure and function which confounds attempts to apply a simple yet comprehensive system of classification.

II. The structure of the coelomic and vascular systems

A. Coelom

Polychaetes have a true coelom which develops from spaces in the embryonic mesoderm (see Chapter 1). In all species it forms an extensive, peritoneum-lined cavity, which may be compartmentalized to a greater or lesser degree by perforate septa. It contains the coelomic fluid and coelomocytes which are circulated by movements of the body wall. The alimentary tract which passes through the coelom is suspended by mesentery and surrounded on its outer surface by peritoneum. The peritoneal lining of the body wall and that surrounding the gut are sometimes distinguished by use of the terms parietal (or somatic) and splanchnic, respectively. The coelom opens to the outside by means of the nephridial openings and in some cases separate genital pores (Fig. 1).

B. Vascular system

The vascular system consists of a series of segmental vessels related to two longitudinal vessels, one of which is dorsal to the gut, in which the blood flows anteriorly, and the other ventral, in which the blood flows posteriorly. Other vessels supply gills when present, and vessels from the gills return blood to the central system (Fig. 1). The main vessels, at least, are contractile and may be specialized as hearts, and contractile blind-ending vessels also occur in some species. The vessels surrounding the gut may be enlarged into a sinus which lies beneath the muscles. The blood vessel walls contain a collagenous connective tissue layer covered by the peritoneum on the outside, with a discontinuous endothelium inside. The endothelial cells often have dendritic extensions and have been described as "sessile blood cells" by early microscopists. The peritoneum contains specialized musculo-epithelial cells and may be enlarged in places to form an "extravasal tissue". It is in this tissue that the plasma haemoproteins are synthesized (Kennedy and Dales, 1958). The dark colour of these cells, which may break free into the coelom is due to haematins formed during haem synthesis (Mangum and Dales, 1965; Potswald, 1969; Dales and Pell, 1970). These cells should not be confused with the "chloragogen" tissues of earthworms. In some

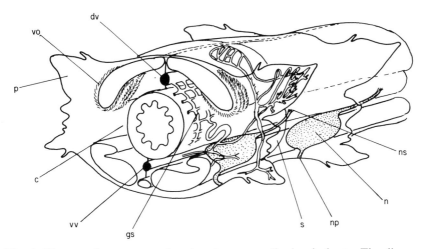

Fig. 1. The vascular system and coelom in a generalized polychaete. The diagram illustrates the main features in *Nereis diversicolor*, based on Goodrich (1893), Lindroth (1938) and Nicoll (1954). Dorsal (dv) and ventral (vv) longitudinal vessels are linked by a gut sinus (gs) with segmental vessels to and from the parapodia (p) and the dorsal surface where respiratory exchange takes place. The coelom (c) drains to the outside by nephridiopores (np) from the paired segmental nephridia (n) which open through the septa (s) at the nephridiostome (ns) in each adjacent segment. In *N. diversicolor*, there are ciliated "vibratile organs" (vo) on which amoebocytes collect, representing coelomostomes. In other polychaetes, septa may be reduced and the nephridia and coelomoducts differently arranged.

polychaetes, there is a "heart-body" which may be regarded as involuted extravasal tissue. The heart-body may also occlude the dorsal vessel in some of the large cirratulids and forms a discrete organ in the ampharetids and terebellids. There is no evidence that the heart-body gives rise to free cells.

The vascular system is typically closed and the plasma haemoproteins are confined to the blood although, in *Nephtys hombergii* and *N. caeca*, haemoglobin (erythrocruorin) is found in the coelomic fluid. Plasma haemoglobins are of high molecular weight ($\simeq 3 \times 10^6$ daltons) molecules being visible under the electron microscope as double hexagons. Haemoglobin molecules can be detected in vesicles within the haemopoietic cells (Potswald, 1969; Dales and Pell, 1970). In a few worms, the vascular system is reduced or has disappeared and the respiratory functions of the blood are assumed by the coelomic fluid. In *Glycera dibranchiata*, *G. americana* and some other species of *Glycera*, the coelom is filled with red cells which function in oxygen transport. Intracellular haemoglobins are of relatively low molecular weight and cannot be detected in electron micrographs. Little work has been carried out on the fine structure of polychaete vessels since Hanson's 1949 review.

III. Structure and classification of coelomocytes and haemocytes

A. Coelomocytes

There is no generally accepted comprehensive classification system for the coelomocytes of polychaetes. Many authors have avoided specific nomenclature, referring to cells by their gross morphological characteristics or employing general terms such as coelomic corpuscles, leucocytes and lymphatic corpuscles. Others have resorted to the use of mammalian terms based upon superficial structural similarities to lymphocytes or granulocytes.

The coelomocytes vary considerably from one species to another but certain types of cell recur throughout the group. The most widely used terms to describe them are "amoebocytes", "eleocytes" and "erythrocytes", according to their functions, but these are not mutually exclusive. Table I shows the distribution of the different coelomocyte types, throughout the polychaete families. Where authors have adopted different terminology, the allocation of cells to one of the three categories is based upon their description. Different types of amoebocyte have not been distinguished. Haemoglobin-containing eleocytes have been entered under both the eleocyte and erythrocyte categories and bracketed to indicate that only one cell type is represented.

1. Amoebocytes

This is by far the largest category and includes a multiplicity of cell types (Fig. 2a). In general, they are small, relatively undifferentiated and potentially phagocytic. They probably occur in all polychaetes, although there are very few in some species and Déhorne (1930b) thought they were absent in *Sabellaria spinulosa*. Liebman (1946), introduced the word "lymphoidocytes" to describe this group of cells, but with the exception of Fretter (1953), and Clark (1961), this term has not been widely adopted.

A common type of amoebocyte is fusiform, filiform or spindle-shaped (Malaquin, 1893; Picton, 1898; Ashworth, 1901, 1904). These may be 50 μm in length with a single, centrally placed, oval nucleus (Fig. 2a) which may appear folded (Fauré-Fremiet, 1925a; Pilgrim, 1965), due to a differentially staining band along its central axis. The cytoplasm characteristically contains bundles of microfilaments (Fig. 2b) which extend throughout the length of the cell. Our own study of the ultrastructure of elongate fusiform cells in *Arenicola marina* shows that their cytoplasm also contains peripheral vacuoles, scattered mitochondria, small pockets of glycogen and occasional granular inclusions (Fig. 2b).

Less attenuated, spindle-shaped cells containing small fibres have been described in *Perinereis cultrifera* (Romieu, 1921a), *Ophelia bicornis* (Tsusaki, 1928), and *Arenicola marina* (Mattisson, 1975). This is a very common cell type in *A. marina* (Figs 2a, 3a,b,c). Their ultrastructure reveals bundles of microfilaments surrounding the nucleus, Golgi complexes and sections of rough endoplasmic reticulum, in addition to other features previously described in the elongated fusiform cells. Déhorne (1924b, 1925) also referred to the perinuclear fibres which occur in *Glycera convoluta* and *Nephtys* cells as the "linome" and called these cells "linocytes" (see oligochaetes, Chapter 4).

Other coelomic amoebocytes, variously described as round, ovoid, spherical, or asteroid, are found throughout the polychaete families, such as the capitellid, *Notomastus profundus* (Picton, 1898), the terebellid, *Polymnia nebulosa* (Picton, 1898), the cirratulid, *Dodecaceria concharum* (Caullery and Mesnil, 1898), the nephtyid, *Nephtys caeca* (Stewart, 1900), the nereid, *Perinereis marioni* (Herpin, 1921), and the aphroditid, *Aphrodite aculeata* (Fordham, 1925). These amoebocytes are relatively small cells, generally 5–20 μm in diameter, capable of producing pseudopodia and frequently described as phagocytic, such as those in *Glycera convoluta* (Goodrich, 1898), *Platynereis dumerilii* (Fretter, 1953), *Perinereis macropus* (Sichel, 1962a), *Amphitrite johnstoni* = *Neoamphitrite figulus* (Dales, 1964), and *Hermodice carunculata* (Marsden, 1966).

Little is known about the structure of these rounded amoebocytes. They

TABLE I. Distribution of polychaete coelomocyte types.

Family/species	Amoebocytes	Eleocytes	Erythrocytes	Citation
APHRODITIDAE				
Aphroditid spp. ⎫	x			Darboux (1900)
Aphrodite aculeata ⎬	x			Romieu (1923)
Lepidonotus squamatus		x		
AMPHINOMIDAE				
Hermodice carunculata	x			Marsden (1966)
PHYLLODOCIDAE				
Eulalia viridis	x			Romieu (1923)
Phyllodoce lamelligera	x	x		Williams (1852)
HESIONIDAE				
Hesione sicula	x			Goodrich (1897)
SYLLIDAE				
Syllis spp. ⎫	x			Malaquin (1893)
Syllis hyalina ⎬				
NEREIDAE				
Nereis fucata	x			Déhorne (1930a)
Nereis diversicolor	x	x		Goodrich (1893)
				Déhorne (1924b)
				Prenant (1929)
				Thomas (1930)
				Dales (1950)
				Herlant-Meewis and Nokin (1963)
Nereis grubei	x	x		Schroeder (1967)
Nereis pelagica		x		Dhainaut (1966)
Nereis succinea	x	x		Baskin (1974)
Nereis spp.	x			Williams (1852)

Species			References
Perinereis cultrifera	x	x	Herpin (1921) Romieu (1921a,b, 1923) Déhorne (1930d) Sichel (1964)
Perinereis marioni *Perinereis macropus* *Platynereis dumerilii* *Micronereis variegata*	x x x x		Herpin (1921) Sichel (1961, 1962a,b, 1964) Fretter (1953) Racovitza (1894)
NEPHTYIDAE *Nephtys* spp.	x		Goodrich (1897) Déhorne (1936)
Nephtys hombergii	x		Williams (1852) Clark (1965)
Nephtys caeca *Nephtys margaritacea*	x x		Goodrich (1904) Cantacuzène (1897)
GLYCERIDAE *Glycera alba* *Glycera americana* *Glycera siphonostoma*	 x x	x x x	Williams (1852) Cowden (1966), Weber and Heidemann (1977) Cuénot (1891, 1897) Goodrich (1898) Kollmann (1908)
Glycera unicornis *Glycera tesselata* *Glycera decipiens* *Glycera convoluta*	x x x	x x x x	Goodrich (1898) Romieu (1921d) Ochi (1969) Déhorne (1924a) Goodrich (1898) Kollmann (1908)
Glycera dibranchiata	x		Hoffman et al. (1968) Shafie et al. (1976) Weber (1978)

TABLE I.—cont.

Family/species	Amoebocytes	Eleocytes	Erythrocytes	Citation
Glycera robusta			x	Terwilliger et al. (1976)
Glycera rouxii			x	Weber and Heidemann (1977) Weber (1978)
Glycera gigantea			x	Andrew (1965), Weber (1978)
Goniada maculata			x	Kollmann (1908)
EUNICIDAE				
Eunice harassi		x		Caullery and Mesnil (1915)
Marphysa sanguinea		x		Prenant (1929)
CIRRATULIDAE				
Dodecaceria concharum	x			Caullery and Mesnil (1898)
Chaetozone setosa	x			Meyer (1887)
Andouinia filigera	x			Picton (1898)
(*Cirriformia tentaculata*)				
Cirratulus cirratus	x			Romieu (1923)
Cirratulus grandis		x	x	Liebman (1946)
SCALIBREGMIDAE				
Scalibregma inflatum	x			Ashworth (1901)
OPHELIIDAE				
Ophelia cluthensis	x			Brown (1938)
Ophelia radiata	x			Schaeppi (1894)
Ophelia bicornis	x			Kunstler and Gruvel (1898)
Travisia forbesii	x	x	x	Romieu (1923) Bloch-Raphael (1939)
Travisia japonica	x		x	Ochi (1969)
Travisia pupa			x	Weber (1978)

Taxon		Reference
CAPITELLIDAE		
Capitella spp.	x	Lankester (1873), Eisig (1887)
		Cuénot (1891)
Dasybranchus caducus		Cuénot (1891)
Notomastus benedeni		Cuénot (1891), Romieu (1923)
Notomastus profundus	x	Eisig (1887), Picton (1897)
Notomastus latericius		Cunningham and Ramage (1888)
		Weber (1978)
Capitella capitata	x	Cunningham and Ramage (1888)
		Wells and Warren (1975)
		Weber (1978)
STERNASPIDAE		
Sternaspis scutata	x	Rietsch (1882)
ARENICOLIDAE		
Arenicola marina	x	Gamble and Ashworth (1900)
		Ashworth (1904)
		Romieu (1923)
		Fauré-Fremiet (1925a,b, 1934)
		Alscher (1949)
		Mattisson (1975)
Arenicola grubei	x	Gamble and Ashworth (1900)
		Romieu (1923)
Arenicola cristata	x	Gamble and Ashworth (1900)
MALDANIDAE		
Clymene lumbricoides	x	Déhorne (1925)
Clymenella torquata	x	} Pilgrim (1965)
Euclymene oerstedi	x	x
AMPHICTENIDAE		
Lagis koreni	x	Déhorne (1925)

TABLE I.—*cont.*

Family/species	Amoebocytes	Eleocytes	Erythrocytes	Citation
TEREBELLIDAE				
Terebellid spp.	x	x		Williams (1852)
Amphitrite ornata	x	x		Liebman (1946)
				Alscher (1949)
				Weber (1978)
Lanice conchilega	x			Déhorne (1922d)
				Romieu (1923)
Polymnia nebulosa	x	x		Picton (1898)
				Siedlecki (1903)
Amphitrite rubra	x	x		Picton (1898)
Amphitrite variabilis	x	x		Romieu (1923)
Amphitrite edwardsi	x	x		Kollmann (1908)
				Romieu (1923)
				Picton (1898)
Neoamphitrite figulus (= *Amphitrite johnstoni*)	x	x	x	Fauré-Fremiet (1928)
				Thomas (1940)
				Dales (1957, 1964)
				Dales and Pell (1970)
Enoplobranchus sanguineus			x	Weber (1978)
Terebella lapidaria	x	x	x	Claparède (1968)
				Romieu (1923)
				Bloch-Raphael (1939)
				Dales (1957, 1964)
				Wells and Dales (1975)

Species			Reference
Terebella ehrenbergi		x	Ochi (1969)
Polycirrus haematodes	x	x	Cuénot (1891)
Pista cristata	x		Romieu (1923)
Pista pacifica		x	Terwilliger and Koppenhoffer (1973), Terwilliger (1974), Weber (1978)
Nicolea zostericola	x		Eckelbarger (1976)
Thelepus crispus		x	Weber (1978)
SABELLIDAE			
Sabella spallanzanii (=*Spirographis spallanzanii*)	x	x	Kollmann (1908), Romieu (1923)
Sabella penicillum	x		Dales (1961)
Megalomma vesiculosum	x		Romieu (1923)
Sabella spp.	x		Williams (1852)
SERPULIDAE			
Serpula vermicularis	x		Prenant (1929)
Protula intestinum	x		
Pomatoceros triqueter	x		

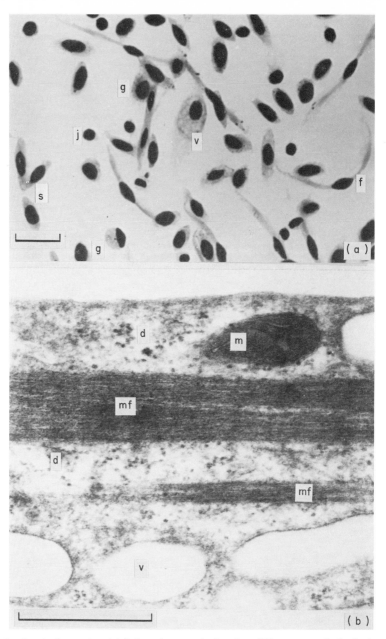

Fig. 2. *Arenicola marina*. (a) light micrograph showing different morphological types of amoebocyte; fusiform (f), spindle-shaped (s), granular (g), juvenile (j) and vacuolated (v). Cytospin preparation stained with Giemsa/May Grünwald. Scale bar represents 20 μm. (b) electron micrograph showing bundles of microfilaments (mf) in a fusiform amoebocyte with characteristic vacuoles (v), mitochondrion (m) and glycogen deposits (d). Scale bar represents 0·5 μm.

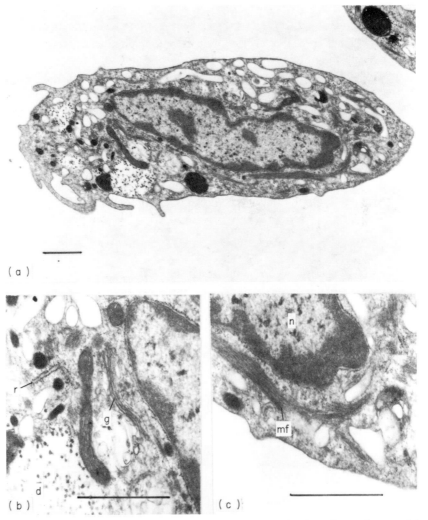

Fig. 3. *Arenicola marina*. Electron micrographs of a spindle-shaped amoebocyte: (a) whole cell; (b,c) enlargements of same cell, (b) showing prominent Golgi body (g), RER (r) and glycogen deposits (d); (c) showing perinuclear microfilaments (mf), nucleus (n). Scale bars represent 1·0 μm.

may be hyaline, such as those in *Perinereis cultrifera* (Romieu, 1921a) and *Eulalia viridis* (Romieu, 1923), though more usually they are granular as in opheliids (Kunstler and Gruvel, 1898; Brown, 1938), and nereids (Romieu, 1921a; Déhorne, 1924b; Sichel, 1961). Both Romieu and Déhorne (loc. cit.) described the granules in *Perinereis cultrifera* cells as eosinophilic, while in *P. macropus*, Sichel (1961) described small, granular phagocytes with eosinophilic cytoplasm containing basophil granules. Yellow or brown, refractile, granular inclusions have been recorded in *Syllis hyalina* (Malaquin, 1893) in *A. marina* (Gamble and Ashworth, 1898; Ashworth, 1904) and in *Aphrodite aculeata* (Fordham, 1925).

More recently, ultrastructural studies have demonstrated the structure of granular amoebocytes in *Nereis succinea* (Baskin, 1974), *Nicolea zostericola* (Eckelbarger, 1976) and *A. marina* (Dixon, unpublished). In *N. zostericola*, these cells may be either amoeboid or fusiform. In all three species, they contain distinctive, homogeneous, electron-dense, cytoplasmic granules which are membrane-bound and rounded or dumb-bell shaped (Figs 4a,b).

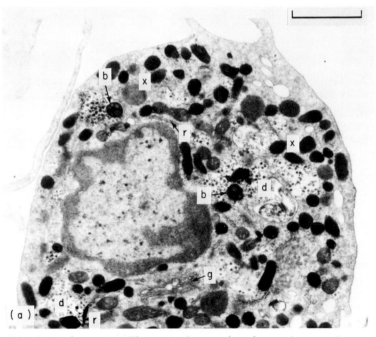

Fig. 4(a). *Arenicola marina*. Electron micrographs of granular amoebocytes; (a) showing prominent membrane-bound granules (x), Golgi body (g), multivesicular bodies (b), glycogen deposits (d) and RER (r), scale bar represents 1 μm;

The cytoplasm also contains Golgi bodies, scattered rough endoplasmic reticulum and glycogen deposits. In *A. marina*, multivesicular bodies are also frequently found (Fig. 4a). In *N. succinea*, Baskin (1974) described the granules as usually being confined with the microtubules, microfilaments and other inclusions, to a central perinuclear region, leaving a peripheral zone of clear cytoplasm containing small vesicles which gives rise to the pseudopodia. Other cells with fewer granules, prominent Golgi bodies and endoplasmic reticulum showed no such pronounced zonation. In *A. marina*, the granular amoebocytes produce hyaline, veil-like pseudopodia and are avidly phagocytic (Figs 4b, 5, 9a).

Small "lymphocyte-like" amoebocytes measuring 5–6 μm in diameter, with a round nucleus surrounded by a thin rim of cytoplasm, have been recorded in *Perinereis cultrifera* (Romieu, 1921a). Similar small amoebocytes are common in *A. marina* (Fig. 2a). In *Nereis diversicolor*, they are said to transform into the oesinophilic, granular amoebocytes (Déhorne, 1924b) and it seems probable also that in other species such cells are juvenile amoebocytes.

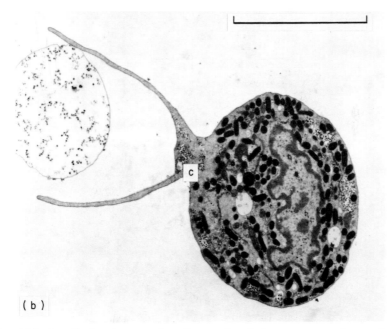

(b)

Fig. 4(b). A cell showing pseudopodia produced from a zone of granule-free cytoplasm (c). Scale bar represents 5·0 μm.

By contrast, large "club" or "bow" cells up to 150 μm in length have been described in *Notomastus profundus* (Picton, 1897), *Ophelia bicornis* (Tsusaki, 1928) and *Ophelia cluthensis* (Brown, 1938). Each cell contains a dark brown, refringent inclusion in the shape of a rod or bow from which the names are derived.

Amoebocytes also become multinucleate in cirratulids (Caullery and Mesnil, 1898; Romieu, 1923; Fauré-Fremiet, 1928).

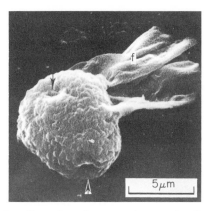

Fig. 5. *Arenicola marina*. Scanning electron micrograph of granular amoebocyte showing fluted, veil-like cytoplasmic extension (f) and surface depressions (un-labelled arrows).

In most instances, the relationship between the different types of amoebocyte is unknown. They may represent discrete cell populations or different developmental stages in a single cell line. In *A. marina*, the appearance of the amoebocytes is modified by their phagocytic activity, and in *P. cultrifera* Romieu (1921a,b) observed that the large eosinophil amoebocyte is unknown. They may represent discrete cell populations or species which also have eleocytes or erythrocytes, the amoebocytes often represent early stages in their development.

Several authors refer to a rapid process by which cells transform from one morphological type to another; from an inactive, passive or resting fusiform condition to an actively amoeboid form. This is heralded in *Clymenella torquata* and *Euclymene oerstedi* (Pilgrim, 1965) by the appearance of basophil granules along the fibrils in the fusiform cells. In coelomic fluid withdrawn from *A. marina*, Fauré-Fremiet (1925a) described the breakdown of the filaments in fusiform cells and the appearance of granules, the cells became globular with petalloid pseudopodia. Cells in this condition *in vitro* stuck to surfaces and to each other to form clumps. The same transformation was described *in vivo* upon the injection of inert particles and represented the beginning of phagocytosis. Similar observations were made by Tsusaki (1928) in *Ophelia bicornis*. While having observed rapid clumping of the coelomocytes upon the removal of coelomic fluid samples from *A. marina*, our own observations show that both fusiform and granular amoebocytes are present and involved in the formation of cellular aggregations.

2. Eleocytes

These are relatively large cells averaging 40 μm in diameter at maturity and are characterized by the presence of nutritive reserves, particularly lipid, in their cytoplasm. The term "eleocyte" was first introduced by Rosa in 1896 to describe lipid-containing cells in oligochaetes, Picton (1898) applying this term to similar cells in the polychaetes *Polymnia nebulosa*, *Amphitrite rubra* and *A. variabilis*. Liebman (1946) described them as "trephocytes" which he regarded as being a cell line separate from the "lymphoidocytes". Once again his terminology has not been widely adopted, although the general term "trephocyte", meaning nutritive cell, might well be more appropriate than "eleocyte" for cells in which the main nutritive material is not fat but glycogen or protein. They have also been referred to as "adipo-spherular cells" (Kollmann, 1908), while Eckelbarger (1976) rather confusingly called them "agranular amoebocytes". In some cases, they contain haemoglobin and may be referred to as erythrocytes, depending on the functional context in which they are being discussed (Fig. 6).

Eleocytes have been described in several polychaete families, particularly the terebellids, nereids and sabellids. In general, they are oval or round, and in *Neoamphitrite figulus* are flattened oval discs (Fig. 6a) (Dales, 1964). Sizes range from 15 μm diameter in *Terebella lapidaria* (Romieu, 1921c) to 70 μm in *Sabella spallanzanii* (Dales, 1961). Measurements of cell size—distribution in individual *Neoamphitrite figulus* (Dales, 1964) suggest that the eleocytes grow rapidly to reach a maximum size characteristic of the species and independent of size, sex or age of the individual. Eleocytes usually have a nucleus but Sichel (1961) observed that this may be absent or pycnotic in *Perinereis macropus*, and Alscher (1949) also claimed that some cells in *Amphitrite ornata* were anucleate. The cytoplasm is invariably granular with accumulations of lipid in the form of globules or granules (Picton, 1898; Romieu, 1921c; Schroeder, 1967) and frequently glycogen (Sichel, 1961; Ochi, 1969; Eckelbarger, 1976). Dales (1961) also described the accumulation of carotenoid pigment in the eleocytes of *Sabella spallanzanii*. In

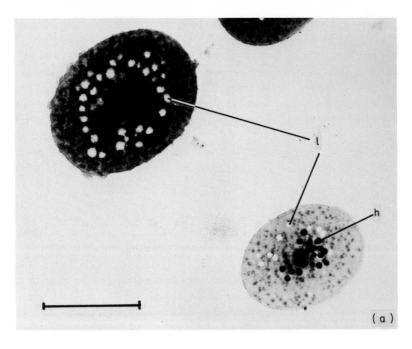

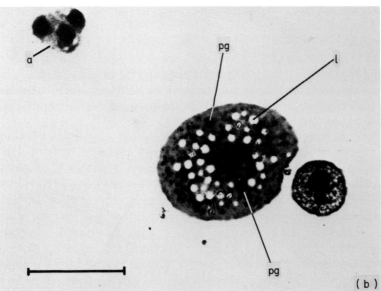

Fig. 6. Light micrographs of eleocytes: (a) *Neoamphitrite figulus* showing lipid (l) and haematin granules (h); (b) *Terebella lapidaria* showing mature eleocyte with lipid (l) and proteinaceous granules (pg), a smaller eleocyte without lipid reserves and a group of three amoebocytes (a). Scale bars represent 25·0 μm. Cells fixed in formaldehyde vapour and stained with solochrome/cresyl fast violet.

Neoamphitrite figulus, certain cells contain large, hyaline spheres which are strongly basophilic but of unknown composition (Dales, 1964).

The presence of large vacuoles in nereid eleocytes described by Romieu (1921b) and Sichel (1961) was confirmed by the ultrastructural studies of Dhainaut (1966) and Baskin (1974) (Fig. 7). In *Nereis succinea*, the vacuoles often contain a flocculent, electron-dense material of unknown composition (Baskin, 1974). In *Nereis pelagica*, the eleocytes are of typical structure, with a large vacuole and abundant reserves in the form of both glycogen and lipid, the endoplasmic reticulum is poorly developed, mitochondria are sparse and the nucleus has no nucleolus. However, in the female epitokous form (the swimming sexual stage), the cells undergo a morphological transformation, characterized by the appearance of a large prominent nucleolus, increased mitochondria and a proliferation of the endoplasmic reticulum (Dhainaut, 1966). A similar switch to protein metabolism occurs in the terebellid, *Nicolea zostericola* at the onset of vitellogenesis (Eckelbarger, 1976), but in this case it is achieved by a changeover in cell production from Type 1 to Type 2 "agranular amoebocytes".

There seems to be little doubt that at least in some species (*Dodecaceria concharum*, Caullery and Mesnil, 1898; *Perinereis cultrifera*, Romieu, 1921b; Sichel, 1964; *P. marioni*, Herpin, 1921; *P. macropus*, Sichel, 1961; *Sabella spallenzanii*, Dales, 1961; *Neoamphitrite figulus*, Dales, 1964), the eleocytes are derived from phagocytic amoebocytes. Intermediate cell types have been described by Herpin (1921), Romieu (1921a,b), and Sichel (1961), and in *Neoamphitrite figulus*, Dales (1964) observed that it is often difficult to distinguish large amoebocytes from small eleocytes. Even large "mature" eleocytes are able to extend pseudopodia and take up particles by phagocytosis, as in *P. cultrifera* (Romieu, 1921b; Sichel, 1962a).

3. Erythrocytes

These are haemoglobin-containing coelomocytes found in the families Terebellidae, Opheliidae, Cirratulidae, Capitellidae and Glyceridae. They are also referred to as "haematocytes" and even "blood cells" but are exclusively coelomic. In the terebellids, opheliids and cirratulids, these cells are haemoglobin-containing eleocytes (Figs 6a,b) but in the glycerids and capitellids, in which the vascular system has been lost, the erythrocytes are specialized solely for transport or storage of oxygen (Weber, 1978).

The more specialized erythrocytes usually take the form of round or oval, nucleated discs (Cuénot, 1891; Picton, 1898; Goodrich, 1898; Dawson, 1933), measuring up to the 20 μm across in *Glycera decipiens* (Ochi, 1969). Ultrastructural studies of these cells in *Glycera americana* (Cowden, 1966), *G. decipiens* (Ochi, 1969) and *G. dibranchiata* (Seamonds and Schumacher,

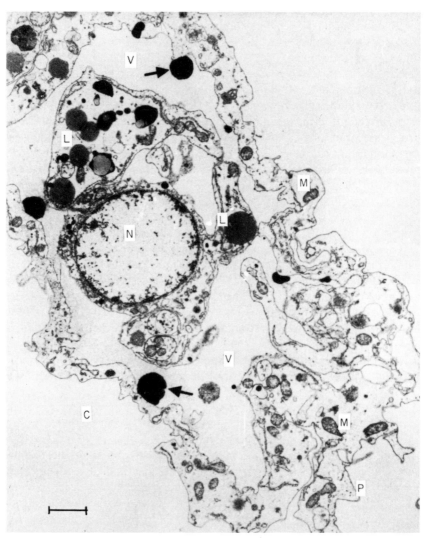

Fig. 7. Electron micrograph of an eleocyte of *Nereis succinea*, showing nucleus (N), lipid droplets (L), mitochondria (M), coelomic fluid (C) and pseudopodia (P). Note electron-dense globules attached to vacuole membrane (unlabelled arrows) scale bar represents 2·0 μm. (From Baskin, 1974).

1972) have shown that the cytoplasm in mature cells is characterized by the sparsity of their organelles. There are, however, scattered mitochondria and sections of smooth endoplasmic reticulum, with haemoglobin distributed throughout the cytoplasm and also located in lysosomes responsible for its breakdown.

In the haemoglobin-containing eleocytes, the lysosomes also contain the by-products of haem synthesis including haematins (Dales, 1964; Mangum and Dales, 1965). The concentration of haemoglobin is variable even from species to species within a genus. The quantity of haemoglobin in *Neoamphitrite figulus* erythrocytes is equal to half of that in the blood and in ageing cells it is converted into haematin.

It is not clear at what stage in the development of the erythrocytes haemoglobin is produced. Cowden (1966) concluded that the absence of both RNA and extensive endoplasmic reticulum indicated that continuous haemoglobin synthesis does not occur in the erythrocytes of *G. americana*. Romieu (1923), had noted earlier that haemoglobin made its appearance in cells of this species when they were only 7–8 μm in diameter. Shafie *et al.* (1976), demonstrated metabolic activity compatible with haemoglobin synthesis in *G. dibranchiata* erythrocytes, and Dales (1968) showed that the coelomocytes of *Neoamphitrite figulus* were able to convert δ-amino laevulinic acid to phosphobilinogen the first step in haem synthesis. Ochi (1969) found that the lysosomes of the cells in *Travisia japonica* gave positive reactions to iron and his figures and descriptions are reminiscent of those described by Dales and Pell (1970) in *N. figulus*, the granules probably representing iron porphyrins accumulated during haemoglobin synthesis (Mangum and Dales, 1965). Kollmann (1908) observed cells in *Glycera siphonostoma* of an intermediate type between the granular ameobocytes and mature erythrocytes.

B. Blood cells (haemocytes)

Even exercising the greatest care it is often impossible to be certain that cells seen in the blood removed from a polychaete are not derived from the coelomic fluid by contamination, especially as coelomocytes often cling to the outside of blood vessels. Previous reference to the presence of abundant haemocytes (e.g. Cuénot, 1891) may be due to such admixture of blood and coelomic fluid, for most observers who report cells in the blood find relatively few. Positive evidence that cells occur within the vessels is nevertheless provided by histological sections and electron micrographs (Dhainaut, 1969a) (Fig. 8). Many "sessile amoebocytes", however, seen within the blood by earlier workers (Vejdovsky, 1882; Liwanow, 1914;

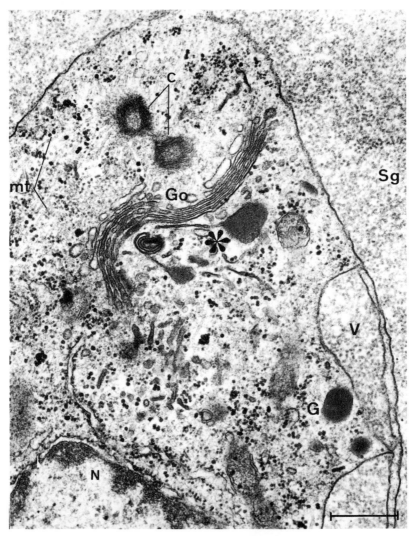

Fig. 8. Haemocyte in *Nereis diversicolor* showing the convergence of microtubules (mt) in direction of the centrioles (c). The saccules on the distal side of the Golgi apparatus (Go) are characterized by the high density of their contents. At their extremity (asterisk) material resembling the granules (G) is elaborated. Note the similarity of the contents of the vacuoles (V) and the blood plasma (Sg). N, nucleus. Scale bar represents 0·5 μm. (From Dhainaut, 1969a).

Probst, 1929) may have been stellate endothelial cells (Hanson, 1951). Clark and Clark (1962), nevertheless, were able to distinguish sessile cells from those freely circulating in the blood in *Nephtys hombergii* since the free cells accumulated near wounds (Clark *et al.*, 1962; Herlant-Meewis and van Damme, 1962). The conflicting accounts of occurrence and absence of cells within the blood vessels of polychaetes, such as *Sabella* sp., may be due to invasions of vessels by amoebocytes from the coelom under certain conditions. Koechlin (1966) and Kryvi (1972) produced ultrastructural evidence for the presence of cells in the vessels in *Sabella penicillum*, though Romieu (1923), Ewer (1941) and Hanson (1951) had earlier maintained that they were absent. Table II shows the distribution of blood cells throughout the polychaete families. Brief descriptions of the cells are given, where available.

We suggest that "haemocyte" should be adopted as a general term to describe cells found free within the blood but, since there is no evidence that they are not of coelomic origin, this does not necessarily imply separate cell lineage.

In general, the haemocytes are small (3–10 μm), nucleate, frequently amoeboid cells which resemble the coelomic amoebocytes which, with the exception of *Magelona papillicornis*, do not contain respiratory pigment. *Nereis diversicolor* haemocytes are fusiform, with microtubules stretching throughout the long axis of the cell (Fig. 8) (Dhainaut, 1969a,b). The nucleus is oval, sometimes lobed, and lacks a nucleolus. The cytoplasm contains electron-dense granules secreted by the Golgi, α and β glycogen, and is permeated by a system of vocuoles, some of which appear to contain material taken in from the surrounding blood and which is undergoing intracellular digestion (Fig. 8). The cytoplasm of *Sabella penicillum* haemocytes appears to be less granular but contains numerous vacuoles and vesicles (Koechlin, 1966; Kryvi, 1972).

IV. Origin and formation of coelomocytes and haemocytes

A. Coelomocytes

In some polychaetes, only one cell-type has been recognized but others have several (Table I), each specialized for different functions. How far these different cell types are separately derived, or whether they simply represent different stages in development of a single or a few cell-lines, is unknown. Many possible sites of production have been suggested, largely based on circumstantial evidence, although some recent histological studies, such as that of Eckelbarger (1976) with the terebellid, *Nicolea zostericola*, provide

TABLE II. Distribution and characteristics of polychaete blood cells (haemocytes).

Family/species	Remarks	Citation
AMPHINOMIDAE *Hermodice carunculata*	rare, identical to coelomocytes	Marsden (1966)
PHYLLODOCIDAE Phyllodocid sp.	—	Cuénot (1891)
SYLLIDAE Syllid sp.	colourless, few	Lankester (1878), Cuénot (1891)
NEREIDAE *Nereis diversicolor*	fusiform, nucleate, granular amoebocytes	Clark et al. (1962), Herlant-Meewis and van Damme (1962) Herlant-Meewis and Nokin (1963), Dhainaut (1969a)
Platynereis dumerilii	numerous, 3 μm, round or fusiform, nucleate, granular	Cuénot (1891)
NEPHTYIDAE *Nephtys hombergii*	—	Clark and Clark (1962) Clark et al. (1962)
EUNICIDAE Eunicid spp.	—	Cuénot (1891)
Eunice siciliensis	sessile	Bergh (1900) Liwanow (1914) } in Hanson (1949)
MAGELONIDAE *Magelona papillicornis*	anucleate globules 2–6 μm, contain haemerythrin	Benham (1897) Déhorne (1932, 1949)
CHAETOPTERIDAE *Chaetopterus* sp.	sessile	Probst (1929), in Hanson (1949)
CIRRATULIDAE Cirratulid spp.	—	Lankester (1878), Cuénot (1891)

CHLORHAEMIDAE		
Flabelligera diplochaetus	sessile	Jourdan (1887), in Hanson (1949)
Trophonia plumosa	few, small, oval nucleate	Cunningham (1888)
OPHELIIDAE		
Ophelia cluthensis	sparse, 7–9 μm, spherical or ovoid nucleate	Brown (1938)
STERNASPIDAE		
Sternaspis scutata	sessile	Vejdovsky (1882) } Goodrich (1904) } in Hanson (1949)
ARENICOLIDAE		
Arenicola spp.	sparse, 4–10 μm, round or ellipsoid, nucleate	Gamble and Ashworth (1900), Ashworth (1904)
Arenicola grubei	sessile	Bergh (1900) } Schiller (1907), } in Hanson (1949) Kermack (1955)
Arenicola marina	—	
MALDANIDAE		
Euclymene oerstedi	6 × 3 μm, non-granular nucleate	Pilgrim (1966)
Clymenella torquata	resemble coelomic phagocytes	
OWENIIDAE		
Owenia fusiformis	sessile	Zurcher (1909), in Hanson (1949)
TEREBELLIDAE		
Terebellid spp.	colourless	Lankester (1878)
Nicolea venustula	numerous, amoeboid, granular, 10 μm	Cuénot (1891)
Polymnia nebulosa	numerous, amoeboid, granular, 10 μm	Cuénot (1891)
Lanice conchilega	sessile	Bergh (1900), in Hanson (1949)
Terebella lapidaria	sessile	Dyrssen (1912), in Hanson (1949)
Amphitrite rubra	sessile	Dyrssen (1912), in Hanson (1949)
SABELLIDAE		
Sabella penicillum	small, nucleate, non-granular	Koechlin (1966), Kryvi (1972)
Amphiglena mediterranea	numerous 6–8 μm, granular, amoeboid	Cuénot (1891)
Sabella spallanzanii	numerous 6–8 μm, granular, amoeboid	Cuénot (1891)
SERPULIDAE		
Serpulid spp.	sessile	Lee (1912), in Hanson (1949)

more convincing evidence. In this species, the agranular amoebocytes appear to originate from specific sites on the lateral parietal peritoneum lining the coelom in all the thoracic and a few of the abdominal segments. It would seem to be a reasonable assumption that, in general, the coelomocytes arise from specialized parts of the peritoneum (Eisig, 1887, capitellids; Gamble and Ashworth, 1898; Kermack, 1955, *Arenicola marina;* Darboux, 1900, polynoids; Fauré-Fremiet, 1928, *Cirriformia tentaculata*; Thomas, 1930, Defretin, 1955, nereids; Dales, 1961, Dales and Pell, 1970, sabellids and *Neoamphitrite figulus*; Marsden, 1966, *Hermodice carunculata*; Pilgrim, 1965, 1966, *Clymenella torquata* and *Euclymene oerstedi*). Some observers have, however, reported that coelomocytes originate from the specialized cells in the epithelial layer surrounding certain blood vessels (Malaquin, 1893, syllids; Darboux, 1900, polynoids; Déhorne and Defretin, 1933b, Dhainaut-Courtois, 1965, Dhainaut, 1969b,c, nereids; Pilgrim, 1965, maldanids).

In *Sabella (Spirographis) spallanzanii*, Cantacuzène (1897) identified a pair of cone-shaped projections near each nephridium, situated against the longitudinal muscles, and suggested that they were a possible source of coelomocytes. In some other polychaetes, distinct "lymph glands" have been described (Fordham, 1926). Cantacuzène (1897) described two pairs of discrete patches of tissue in *Glycera convoluta*, one dorsal and one ventral on the coelomic peritoneum, which were possible sites of coelomocyte production, but without positive proof that coelomocytes did arise there. Similarly, Goodrich (1898) and Kollmann (1908), later expressed doubts in relation to *Glycera* sp. as to whether cells in the nephridial caecum were produced there, as Meyer (1887) maintained, or as seems more likely they accumulated there from the coelomic fluid. It is still not clear from more recent studies (Cowden, 1966; Seamonds and Schumacher, 1972) where the coelomocytes of *Glycera* spp. arise.

In *Arenicola piscatorium*, Cantacuzène (1897) identified masses of amoebocytes close to the nephridia and small masses attached to the ventral longitudinal muscles in the non-nephridial segments, but it is not clear, whether these masses of cells represent accumulations rather than sites of production. Déhorne (1922a) and Déhorne and Defretin (1933b) concluded that, in *Nereis pelagica*, the coelomocytes arose from cells attached to the ventral vessel and that masses of cells seen elsewhere were accumulations of free cells. Nevertheless, it is possible that some actively phagocytic cells may never become detached from the coelomic epithelium. On the other hand, coelomic amoebocytes attach themselves to and creep very readily on suitable surfaces, and it seems more likely that the "vibratile organs" of *Nereis* are areas of accumulation rather than of production (Goodrich, 1893).

Once released from the peritoneum into the coelom, cells may change morphologically but subsequent division probably does not occur as mitoses have rarely been seen in free coelomocytes.

Infection, wounding or transection of the body, which might stimulate cell production in response either to replacement of cells lost or in response to enhanced need for defence, do not appear to initiate division. Our observations on *Nereis diversicolor*, *Terebella lapidaria* and *A. marina* using both ³H-thymidine and nuclear stains have failed to detect cell division in free cells. Fauré-Fremiet (1934) came to the same conclusion from his observations on *A. marina* and *Nephtys hombergii*. Romieu (1923), on the other hand, occasionally saw nuclear division in *T. lapidaria* cells, but he expressed some doubt as to whether cytoplasmic division followed. Déhorne (1930a) also described mitosis in *Nereis fucata* amoebocytes in response to parasitic infection and Kollmann (1908) illustrated amitotic division in *G. convoluta* erythrocytes. Ochi (1969) also reported dividing nuclei in *G. decipiens* cells, and Kunstler and Gruvel (1898) believed that the "red cells" of *Ophelia* sp. reproduced by division on the coelomic fluid. It is, therefore, unwise to generalize about the origins of coelomocytes in the light of these different observations.

B. Blood cells (haemocytes)

In *Magelona papillicornis*, Buchanan (1895) described a syncytial mass, present in the larva, at the posterior end of the dorsal vessel. This mass apparently broke up into fragments, which later divided and lost their nuclei to form haemerythrin-containing corpuscles (Benham, 1897; Déhorne, 1932, 1949). In the adult, these are derived from specialized regions of the dorsal vessel walls. In all other polychaetes, haemocytes are always nucleated and resemble the coelomic amoebocytes which may migrate into the blood vessels. Their origin remains obscure.

V. Functions

A. Coelomocytes

Many coelomocytes are phagocytic and able to migrate through the body tissues to perform diverse functions, including the absorption, digestion and distribution of food materials and the segregation or removal of waste products, degenerate "self" components and invading foreign organisms. They may also participate in wound closure, and some function as erythrocytes by virtue of their haemoglobin content.

1. Nutritive functions

Kermack (1955) and Marsden (1966) have described the role of wandering amoebocytes in relation to digestion in *Arenicola marina* and *Hermodice carunculata*. In *A. marina*, amoebocytes are involved in the intracellular digestion and distribution of nutritive materials. Epithelial cells of the stomach and glandular region of the oesophagus engulf food particles which are subsequently taken up by wandering amoebocytes. These either pass into the blood system or cross the gut wall to the peritoneum, digest their contents and distribute the products through the blood system or coelomic fluid. Kermack (1955) suggested that intracellular digestion by the amoebocytes, in addition to extracellular digestion in the gut, maximizes the extraction of food from the large quantities of indigestible material in the animals' diet.

In *H. carunculata*, Marsden (1966, 1968a) demonstrated esterase activity in amoebocytes which invade the anterior intestinal epithelium, some hours after feeding, and also in coelomocytes in the body cavity. Acid and alkaline phosphatase activity was also demonstrated in coelomocytes associated with peritoneal tissues, blood vessels and the walls of the digestive tract. Marsden (1966, 1968a) suggested that the coelomocytes may contribute to the digestion and absorption of nutrients and may transfer lipid from the intestine to fat stores on the peritoneum. Dales (1957) also suggested that in *A. marina* and *Nereis diversicolor* amoebocytes may transfer surplus fat from the gut to the body wall and glycogen to the peritoneum.

In polychaetes which have eleocytes or "trephocytes", these cells build up stores of metabolic reserves, chiefly lipid and glycogen. We do not know how the eleocytes derive their stores, whether this is by absorption from blood vessels when in contact with them, directly from the gut or by endocytosis from the coelomic fluid. In *Nereis pelagica* and *Nicolea zostericola*, trephocytes synthesize protein for export (Dhainaut, 1966; Eckelbarger, 1976). In all cases, it is probable that these reserves are destined to be utilized in gamete production.

Cell abundance has been shown to vary according to sexual development in a number of species. In *Dodecaceria concharum*, the number increases until the appearance of the gametes and then decreases and finally disappears in the female at maturity (Caullery and Mesnil, 1898). In *Eunice harassi*, Caullery and Mesnil (1915) state that the coelomocytes become more numerous during gametogenesis, and Malaquin (1893) also noted that in syllids the number of cells varies with the season. Dales (1964) recorded that the quantity of cells in *Neoamphitrite figulus* (*Amphitrite johnstoni*), estimated by packed centrifuged volumes and also by weighing, was directly related to total body weight; a worm weighing 0·3 g (dry weight) would contain 0·5 ml packed coelomocytes which represented approximately 0·14 g (dry weight), more than a third of the total dry weight of the body. Seasonal estimations suggest that the number of coelomocytes increases during gamete production and decreases at maturity, implying an annual cycle of production. In *N. zostericola*, Eckelbarger (1976) recorded an increase in the number of coelomocytes during early oogenesis with a change within the cell population from carbohydrate and lipid storage to protein production during vitellogenesis. *N. diversicolor* becomes so filled with coelomocytes during gamete maturation that the coelom may be packed with a kind of "parenchymatous tissue" which disappears at maturity (McIntosh, 1907; Dales 1950).

The nutritive function of coelomocytes in relation to gamete maturation has also been reported by Claparède (1868, 1874) and Cuénot (1891) in *Sabella spallanzanii*, *Cirriformia tentaculata*, *Nereis fucata* and *Psygmobranchus protensis;* Romieu (1921a,b) and Herpin (1921) in *Perinereis cultrifera*; Caullery and Mesnil (1915) in *Eunice harassi*; Dales (1961) and Pocock *et al.* (1971) in sabellids; and particularly by Schröder (1886), McIntosh (1907) and Dales (1950, 1957) in *N. diversicolor*, and most recently at the ultrastructural level by Dhainaut (1966, 1970a,b), Shroeder (1967) and Baskin (1974). Quantitative determinations of lipid and glycogen in the coelomocytes of *Neoamphitrite figulus* (Dales, 1957) indicate that these cells constitute a considerable proportion of the total body content of these reserves and that their principal functions is to supply nutrients to the gametes. In *Sabella spallanzanii*, measurements of these reserves in the coelomocytes and gametes at different seasons demonstrated that the quantities accumulated by the eleocytes during early gametogenesis are roughly equivalent to those found in the mature oocytes (Dales, 1961), again suggested that the eleocyte reserves are destined for the gametes. It is probable that the accumulation of reserves in relation to reproduction is under hormonal control (Clark, 1964).

There appears to be some variation in the mechanisms of transferring nutrients to the growing gametes. Clark and Olive (1973), demonstrated that the gametes arise from distinct sub-peritoneal structures and while some gametes mature with the aid of special nurse cells, as in *Diopatra cuprea* (Anderson and Huebner, 1968), others elaborate their yolk from nutrients absorbed from the coelomic fluid (King *et al.*, 1969; Eckelbarger, 1975). Several authors have reported close proximity or cellular contact between eleocytes and oocytes (Dales, 1964; Dhainaut, 1966; Shroeder, 1967; Potswald, 1972; Eckelbarger, 1975, 1976) but there is no morphological evidence of direct nutrient transfer between them.

2. Phagocytosis and encapsulation

There are abundant references to the phagocytosis by coelomocytes of small particles such as powdered carmine, Indian ink or carbon (Kowalewsky, 1889, 1895 and Metchnikoff, 1893, *Terebella*; Goodrich, 1900, *Nephtys caeca*; Goodrich, 1919, *Arenicola marina*; Sichel, 1962a, *Perinereis macropus*), while larger particles become encapsulated (Tsusaki, 1928, *Ophelia bicornis*). *A. marina* amoebocytes avidly ingest mammalian erythrocytes and bacteria such as *Micrococcus lysodeikticus* both *in vitro* and *in vivo* (Figs 9a,b). These cells also stain strongly for acid-phosphatase activity but tests for β-glucosaminidase and N-acetyl-β-hexosaminidase have given negative results (Dixon, unpublished).

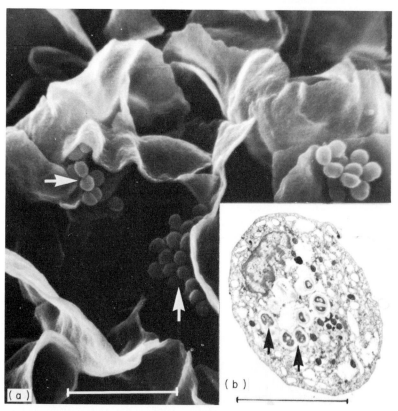

Fig. 9. *Arenicola marina* phagocytes; (a) scanning electron micrograph of a group of phagocytes engulfing clusters of *Micrococcus lysodeikticus* (unlabelled arrows). (b) electron micrograph of phagocyte showing ingested *M. lysodeikticus* enclosed in vacuoles (unlabelled arrows). Scale bars represent 5·0 μm.

Coelomocytes also react to other organisms such as gregarine protozoa either by phagocytosis or by encapsulation (Labbé and Racovitza, 1897, *Leiocephalus leiopygos*; Caullery and Mesnil, 1898, *Dodecaceria concharum*; Siedlecki, 1903, *Polymnia nebulosa*; Goodrich, 1919, *A. marina*; Fauré-Fremiet, 1928, *Cirriformia tentaculata*; Tsusaki, 1928, *Ophelia bicornis*; Déhorne, 1930a,c, *Nereis diversicolor*; Hentschel, 1930, *Arenicola caudata*; Smith, 1950, *Nereis limnicola*). Brasil (1904), described in some detail the reaction of *Pectinaria koreni* cells to the gregarine, *Urospora lagidis*, concluding that the amoebocytes encapsulated the sporozoites only when the parasites were developing. The mature sporozoites are apparently normally too active to allow encapsulation to occur (Siedlecki, 1903), although Cuénot (1901) thought that the cells were actively repelled by the trophozoites. In cirratulids parasitized by *Gonospora longissima*, the amoebocytes are able to destroy the living trophozoites (Caullery and Mesnil, 1898), but *Diplocystis* sp. appears to be attacked only when dead. Siedlecki (1903) found in *Polymnia nebulosa* that while the gregarine *Caryotropha* sp. was encapsulated, flagellates were not attacked. Gastrotrichs which invade the coelom of *Euclymene oerstedi* and *Clymenella torquata* are also both ingested and digested by the coelomocytes (Pilgrim, 1965).

Although the digenean trematode, *Zoogonius rubellus*, artificially introduced into *Nereis virens* failed to develop (Stunkard, 1943), this may have been due simply to an unsuitable *milieu* and there is no evidence that polychaetes can acquire immunity to parasites (Margolis, 1971).

Allen (1904), remarks that parasitic trematodes in the body cavity of *Poecilochaetus* sp. are always encapsulated and Caullery and Mesnil (1899) recorded the uptake of metacercarian cysts by the coelomocytes of *Dodecaceria concharum*. Carton (1967) has also described the ejection of an experimentally introduced parasitic copepod from the body wall of *Sabella penicillum* and *S. spallanzanii* by the formation of a cyst built up by the coelomocytes around the parasite.

Degenerating "self" components are also destroyed by the coelomocytes, such as the muscle cells which undergo histolysis in nereids during metamorphosis to the swimming epitokous form (Déhorne, 1922a,b,c, 1934; Déhorne and Defretin, 1933a,b; Defretin, 1949; Clark, 1961). In nereids which do not have an epitoke, the muscles of the body wall still break down to some degree and are phagocytozed (Herpin, 1925; Déhorne, 1934). It appears to be the eleocytes which are responsible for muscle clearance and this has been recorded at the ultrastructural level by Dhainaut (1966) in *Nereis pelagica* and Baskin (1974) in *N. succinea* (Fig. 10).

After the spawning period the amoebocytes also clear the coelom of redundant gametes and disintegrating remnants of gonadial tissue. At least in some populations of *Nereis diversicolor*, the sexes are unequally represented and as this species is atokous there are many redundant females in the late spring. Thomas (1930), Déhorne (1934) and Dales (1950) have observed amoebocytes phagocytozing degenerating oocytes, and the same phenomenon was reported by Downing (1911) in *Arenicola cristata*. In *Euclymene oerstedi* and in *C. torquata*, Pilgrim (1965) demonstrated peroxidase activity in cells which ingested and digested the left-over gametes and associated debris. Siedlecki (1903), recorded the phagocytosis of redundant sperm and sperm plates in *Polymnia nebulosa*, while in the same species Kowalevsky (1889) injected sperm from other individuals but found no amoebocyte response, although the sperm cells were cleared by the phagocytic action of the nephridial cells.

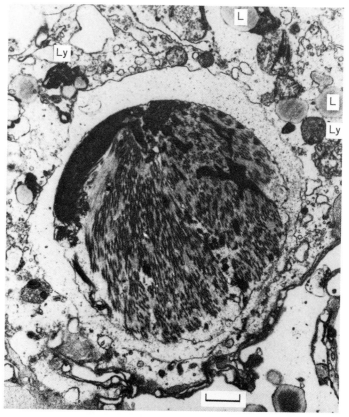

Fig. 10. Electron micrograph of a muscle fibre (sarcolyte) engulfed by an eleocyte in *Nereis succinea*. Note lysosome-like inclusion (Ly) and lipid droplets (L). Scale bar represents 1·0 µm. (From Baskin, 1974).

Thomas (1930), tested the reaction of *N. diversicolor* coelomocytes to various substances, including degenerating oocytes, and claimed that these stimulated the production of cells. Inoculation with homogenized sea-urchin eggs (*Paracentrotus* sp.) evoked no response, though a marked response was found when *Bacterium tumifaciens* was injected into worms in which cell production had apparently been stimulated by oocyte degeneration, large tumour-like growths resulting (Thomas, 1932). Smith (1950), noticed in his study of the viviparous *Nereis limnicola* that deteriorating oocytes and larvae were invested by coelomocytes. Similarly, in worms undergoing histolysis, only free muscle cells were engulfed, indicating that the phagocytes are able to respond to some product of degeneration or a change in the characteristics of their own tissue. Little is known about the factors to which amoebocytes respond and the cellular basis for recognition is unknown.

There are many assertions in the older literature that coelomocytes contain material of excretory origin (Schaeppi, 1894, *Ophelia radiata*; Cantacuzène, 1897, *Sabella spallanzanii*; Caullery and Mesnil, 1898, *Dodecaceria concharum*; Picton, 1897, 1898, *Notomastus profundus* and *Cirriformia tentaculata*). Amoebocytes containing metabolic waste materials and indigestible products of phagocytic activity may be voided from the body by a number of routes or sequestered in excretory depots where they do not interfere with the activities of the worm. In some cases, the cell contents alone may be eliminated. Fage (1906), suggested that the nephridia were responsible for the removal of waste material from the coelom, and in *Nephtys caeca*, following intracoelomic injection with carmine or ink particles, Stewart (1900) described the accumulation of loaded phagocytes around the nephridia, through which the particles were expelled. Schneider (1899), was of the opinion that in terebellids and amphictenids the nephridial cells were also phagocytic. In *A. marina*, the digestive amoebocytes return to the gut lumen to deposit indigestible remains which then pass out of the body with the faeces (Kermack, 1955). Phagocytic masses may also accumulate in the coelom, perhaps forming "brown bodies" (Kermack, 1955), while in *Euclymene oerstedi* and *Clymenella torquata*, phagocytes loaded with inclusions migrate to the tail which is periodically autotomized (Pilgrim, 1965). Particles are also deposited by phagocytes in the epidermis (Racovitza, 1895, *Leiocephalus leiopygos*; Eisig, 1887, *Capitella* sp.). Fretter, (1953) found that following uptake by *Platynereis dumerilii* of ^{90}Sr, labelled amoebocytes migrated through the epidermis to the outer surface of the body. In *Hermodice carunculata*, Marsden (1966, 1968b) has described a mid-ventral line which is probably an excretory device fed by wandering amoebocytes. Amoeboid cells containing granules of various sorts are incorporated into tracts of waste material in the pharyngeal walls. This material is carried forward and either discharges into the buccal cavity or forms a mid-ventral line on the outer surface of the body which is continually shed.

3. Wound closure and regeneration

Participation in would repair seems to be restricted to the initial accumulation at the wound of a plug or "cicatrix" of coelomocytes, effectively closing the wound to prevent excessive loss of fluid and cells, and precluding the entry of microorganisms (Dales, 1978). Following tail amputation in *Nephtys* sp, Clark and Clark (1962) noted a slow migration of coelomocytes to the wound 6–12 h after injury, damaged tissue being removed and accumulating in fixed phagocytes around the nephridia and parapodial blood vessels.

Needham (1970) has reviewed what is known concerning haemostatic mechanisms in polychaetes, and while drawing attention to cell agglutination, points out that cells forming the wound plug are mutually attracted and attach themselves to each other by their pseudopodia. Fauré-Fremiet (1925a), confirmed that in *A. marina* there was no "coagulum" or precipitated network of fibrin in the wound in contrast to that reported by Gamble and Ashworth (1898, 1900), Ashworth (1904), and Romieu (1923), and that the cells forming the scar tissue attached by means of their pseudopodia (Goodrich, 1919). Siedlecki (1903), came to the same conclusion from his observations on *Polymnia nebulosa*, and our own observations on *A. marina* and *N. diversicolor* confirm this. *A. marina* coelomocytes aggregate spontaneously to form large clumps as soon as they are removed from the body. The factors effecting this change in adhesiveness are not known but Fauré-Fremient (1934), maintained that the cells were transformed in response to u.v. light. This transformation could be slowed by low temperatures, and prevented by reducing the pH. Although the cells are reported to disaggregate to some extent under optimal *in vitro* conditions (Bottazi, 1902; Romieu, 1923; Fauré-Fremiet, 1925a, 1934), the cells at the centre of the clumps appear to lose their integrity, breaking down and releasing pigmented granules. Clark (1965), noticed that the pH of the coelomic fluid rose near the wound when the tail was amputated in *Nephtys* sp. and suggested, first, that the coelomocytes might migrate along a pH gradient, and, secondly, that cell adhesiveness may also be affected by the change in pH.

During regeneration of lost parts, large numbers of cells may accumulate near the area of tissue repair (Faulkner, 1932; Liebman, 1946; Clark and Clark, 1962; Herlant-Meewis and Nokin, 1963; Clark, 1965; Pilgrim, 1965), but there is no evidence that these cells participate in tissue removal. In *Nephtys* sp., Clark and Clark (1962) and Clark (1965) found that the coelomocytes migrate from only a few adjacent segments to a wound in order to close it. The cells lose their original character and become more like dedifferentiated mesenchyme. Herlant-Mewis and van Damme (1962) confirmed that in *N. diversicolor*, the first stage of regeneration after tail amputation is an accumulation of amoebocytes which do not subsequently contribute to the regenerated tissue which is derived from the established tissue adjacent to the wounded area. Thouveny (1967), on the other hand, maintained that in *Owenia fusiformis*, the coelomocytes give rise to the coelomic mesoderm in anterior regeneration.

Thus, the coelomocytes do not normally appear to act as "neoblasts" during regeneration (Clark and Clark, 1962; Herlant-Meewis and Nokin, 1963; Boilly, 1969), and contribute to wound repair only by initial plugging to prevent loss of fluid and cells. Such cells are themselves subsequently removed by other phagocytes.

4. Respiration

While the occurrence of coelomic erythrocytes is quite widespread (Table I), their precise role in oxygen transport has been studied in relatively few species, mainly glycerids and terebellids, whose cells may contain high concentrations of haemoglobin. In some terebellid species, however, the haemoglobin concentration in the coelomic erythrocytes may be quite low, possibly serving only to transfer oxygen from the coelomic fluid to the cell itself which is metabolically active during early maturation. In *Glycera* spp., the functions of the blood vascular system have been replaced by the coelom and the coelomic erythrocytes. Erythrocyte function has been reviewed by Mangum *et al.* (1975) and Weber (1978).

B. Blood cells (haemocytes)

Since the cells found within the blood vessels are generally described as small amoebocytes, it seems reasonable to assume that they remove unwanted material from the vessels. This view is supported by the ultrastructure of *Nereis diversicolor* haemocytes, in which particles from the blood appear to be taken up into cytoplasmic vacuoles and subsequently degraded (Dhainaut, 1969a). Concentric, electron-dense lamellae, present in some of the vacuoles, are interpreted as possible products of haemoglobin breakdown, and the abundant deposits of glycogen observed in the cytoplasm might also be derived from the blood.

The participation of haemocytes in wound closure and repair has not been demonstrated but they may function in a similar way to the coelomic amoebocytes. Clark *et al.* (1962) observed in *Nephtys hombergii* a ten-fold increase in the number of haemocytes in vessels near to a wound, whilst in *N. diversicolor*, following caudal amputation, Herland-Meewis and van Damme (1963) observed a large number of "pro-leucocytes" throughout the circulatory system, especially in blood vessels near to the wound. Whether such local increases in the numbers of haemocytes indicates proliferation or simple accumulation—active or passive—is unknown.

VI. Summary and concluding remarks

Polychaete free cells are notable for their variety of structure and function. We still know very little about their ultrastructure, reactions, origin or function, except in one or two species and even in them only from some viewpoints. Whether blood cells are different from coelomocytes remains in doubt. The free cells in the coelom which may be sparse or extremely abundant, according to species, may be ascribed to one of three categories: (1) amoebocytes, some of which at least are demonstrably phagocytic; (2) eleocytes, whose function is mainly one of food storage, probably in relation to gamete production; (3) coelomic erythrocytes, which function in oxygen transport. There is no unequivocal evidence that these types represent separate cell lines. In some worms, the coelom may contain several different types of coelomocyte. In others, the characteristics of each category may be shared or there may be intermediate cells which could be ascribed to any of these three categories. While storage eleocytes or coelomic erythrocytes are characteristic of particular polychaete families or of particular species, all polychaetes that have been carefully examined possess coelomic amoebocytes which probably function as phagocytes, maintaining the body in an aseptic condition and removing dead or moribund cells.

Acknowledgements

We would like to thank Mr Z. Podhorodecki and Mr R. Jones, both of Bedford College, for assistance with photography and electron microscopy.

References

Allen, E. J. (1904). *Quart. Jl. Microsc. Sci.* **48**, 79–151.
Alscher, R. P. (1949). *Biol. Bull., Woods Hole* **97**, 253–254.
Anderson, E. and Huebner, E. (1968). *J. Morph.* **126**, 163–198.

Andrew, W. (1965). "Comparative Hematology". Grune and Stratton, New York and London.

Ashworth, J. H. (1901). *Quart. Jl. Microsc. Sci.* **45**, 13–15.

Ashworth, J. H. (1904). *Arenicola.* L.M.B.C., Memoir 18, *Proc. Liverpool Biol. Soc.*, **18**, 209–326.

Baskin, D. G. (1974). *In* "Contemporary Topics in Immunobiology" (E. L. Cooper, Ed.), Vol. 4, pp. 55–64. Plenum Press, New York.

Benham, W. B. (1897). *Quart. Jl. Microsc. Sci.* **39**, 1–17.

Bergh, R. S. (1900). *Mitt. Hist-Emb. Inst. Univ. Købn.* **49**, 598–624.

Bloch-Raphael, C. (1939). *Ann. Inst. Océanogr. Paris* **19**, 1–78.

Boilly, B. (1969). *Arch. Zool. exp. gén.* **110**, 127–143.

Bottazi, F. (1902). *In* "Handbuch der Vergleichenden Physiologie" (H. Winterstein, Ed.), Vol. 1, pp. 461–596. Fisher, Jena.

Brasil, L. (1904). *Arch. Zool. exp. gén.* (4) **2**, 91–255.

Brown, R. S. (1938). *Proc. Roy. Soc. Edinb.* **58**, 135–160.

Buchanan, F. (1895). *Rep. Brit. Ass. Adv. Sci.* **65**, 469–470.

Cantacuzène, J. (1897). *C. r. hebd. Acad. Sci. Paris* **125**, 326–328.

Carton, Y. (1967). *Arch. Zool. exp. gén.* **108**, 387–411.

Caullery, M. and Mesnil, F. (1898). *Ann. Univ. Lyon* **39**, 1–200.

Caullery, M. and Mesnil, F. (1899). *C. r. hebd. Séanc. Acad. Sci. Paris* **128**, 457–460.

Caullery, M. and Mesnil, F. (1915). *Soc. Biol., Paris* **78**, 593–596.

Claparède, E. (1861). *Reichert's Arch. nat.* **1861**, 542–544.

Claparède, E. (1864). *Mem. Soc. Phys. Hist. nat. Genève* **17**, 465–600.

Claparède, E. (1868). *Mem. Soc. Phys. Hist. nat. Genève* **19**, 313–584.

Claparède, E. (1874). *Mem. Soc. Phys. Hist. nat. Genève* **22**, 1–200.

Clark, M. E. and Clark, R. B. (1962). *Zool. Jb. (Zool.)* **70**, 24–90.

Clark, R. B. (1961). *Biol. Rev.* **36**, 199–236.

Clark, R. B. (1964). *Gen. comp. Endocr.* **4**, 82–90.

Clark, R. B. (1965). *Ann. Rev. Oceanogr. Mar. Biol.* **3**, 211–255.

Clark, R. B. and Olive, P. J. W. (1973). *Ann. Rev. Oceanogr. Mar. Biol.* **11**, 176–223.

Clark, R. B., Clark, M. E. and Ruston, R. J. G. (1962). *In* "Neurosecretion" (H. Heller and R. B. Clark, Eds), pp. 275–286. Academic Press, New York.

Cowden, R. (1966). *Trans. Amer. Microsc. Soc.* **85**, 45–53.

Cuénot, L. (1891). *Arch. Zool. exp. gén.* (2) **9**, 410–447, 667–668.

Cuénot, L. (1897). *Archs Anat. microsc.* **1**, 153–192.

Cuénot, L. (1901). *Archs Biol.* **17**, 581–649.

Cunningham, J. T. (1888). *Quart. Jl. Microsc. Sci.* **28**, 239–277.

Cunningham, J. T. and Ramage, G. A. (1888). *Trans. Roy. Soc. Edinb.* **33**, 635–684.

Dales, R. P. (1950). *J. mar. biol. Ass. U.K.* **29**, 321–360.

Dales, R. P. (1957). *J. mar. biol. Ass. U.K.* **36**, 91–110.

Dales, R. P. (1961). *Quart. Jl. Microsc. Sci.* **102**, 327–346.

Dales, R. P. (1964). *Quart. Jl. Microsc. Sci.* **105**, 262–279.

Dales, R. P. (1968). *Nature, Lond.* **217**, 553.

Dales, R. P. (1978). *In* "Physiology of Annelids" (P. J. Mill, Ed.), pp. 479–508. Academic Press, London and New York.

Dales, R. P. and Pell, J. S. (1970). *Z. Zellforsch. mikrosk. Anat.*, **109**, 20–32.

Darboux, J. G. (1900). *Bull. Soc. Sci. nat. Nîmes* **27**, 53–58.

Dawson, A. B. (1933). *Biol. Bull., Woods Hole* **64**, 233–242.

Defretin, R. (1949). *Ann. Inst. océanogr. Monaco* **24**, 117–257.

Defretin, R. (1955). *Arch. Zool. exp. gén.* **92**, 72–140.

Déhorne, A. (1922a). *C. r. hebd. Séanc. Acad. Sci. Paris* **174**, 1043–1045.
Déhorne, A. (1922b). *C. r. hebd. Séanc. Soc. Biol.* **87**, 1305–1307.
Déhorne, A. (1922c). *C. r. hebd. Séanc. Acad. Sci. Paris* **174**, 1299–1301.
Déhorne, A. (1922d). *C. r. hebd. Séanc. Soc. Biol.* **87**, 1307–1308.
Déhorne, A. (1924a). *C. r. hebd. Séanc. Acad. Sci. Paris* **178**, 1431–1435.
Déhorne, A. (1924b). *C. r. hebd. Séanc. Soc. Biol.* **91**, 303–304.
Déhorne, A. (1925). *C. r. hebd. Séanc. Acad. Sci. Paris* **180**, 333–335.
Déhorne, A. (1930a). *C. r. hebd. Séanc. Soc. Biol.* **104**, 490–493.
Déhorne, A. (1930b). *C. r. hebd. Séanc. Soc. Biol.* **104**, 647–649.
Déhorne, A. (1930c). *C. r. hebd. Séanc. Soc. Biol.* **103**, 665–668.
Déhorne, A. (1930d). *C. r. hebd. Séanc. Soc. Biol.* **103**, 663–665.
Déhorne, A. (1932). *C. r. hebd. Séanc. Acad. Sci. Paris* **195**, 79–81.
Déhorne, A. (1934). *C. r. hebd. Séanc. Acad. Sci. Paris* **199**, 231–233.
Déhorne, A. (1936). *Mem. Mus. Hist. nat. Belg.* (2), **3**, 679–700.
Déhorne, A. (1949). *Arch. Zool. exp. gén.* **86**, 41–49.
Déhorne, A. and Defretin, R. (1933a). *C. r. hebd. Séanc. Soc. Biol.* **113**, 674–676.
Déhorne, A. and Defretin, R. (1933b). *C. r. hebd. Séanc. Soc. Biol.* **113**, 677–680.
Dhainaut, A. (1966). *C. r. hebd. Séanc. Acad. Sci. Paris* **262**, 2740–2743.
Dhainaut, A. (1969a). *C. r. hebd. Séanc. Acad. Sci. Paris* **268**, 711–712.
Dhainaut, A. (1969b). *J. Microscopie* **8**, 69–86.
Dhainaut, A. (1969c). *Z. Zellforsch. mikrosk. Anat.* **96**, 75–86.
Dhainaut, A. (1970a). *J. Microscopie* **9**, 99–118.
Dhainaut, A. (1970b). *Z. Zellforsch. mikrosk. Anat.* **104**, 375–389.
Dhainaut, A. (1976). *In* "Actualités sur les hormones invertebrés", *Colloq. Internat. CNRS*, p. 251.
Dhainaut-Courtois, N. (1965). *C. r. hebd. Séanc. Acad. Sci. Paris* **261**, 1085–1088.
Downing, E. (1911). *J. Morph.* **22**, 1001–1043.
Dyrssen, A. (1912). *Jena Zeits. Naturw.* **48**, 365–398.
Eckelbarger, K. J. (1975). *Mar. Biol.* **30**, 353–370.
Eckelbarger, K. J. (1976). *Mar. Biol.* **36**, 169–182.
Eisig, H. (1887). *Fauna und Flora von Neapel* **16** (1–2).
Ewer, D. W. (1941). *Quart. Jl. Microsc. Sci.* **82**, 587–620.
Fage, L. (1906). *Ann. Sci. nat.* (9) **3**, 261–410.
Faulkner, G. H. (1932). *J. Morph.* **53**, 23–58.
Fauré-Fremiet, E. (1925a). *C. r. Ass. Anat. Paris* **20**, 226–232.
Fauré-Fremiet, E. (1925b). *C. r. hebd. Séanc. Acad. Sci. Paris* **180**, 396–399.
Fauré-Fremiet, E. (1928). *Bull. Biol. Fr. Belg.* **62**, 149–156.
Fauré-Fremiet, E. (1934). *Arch. exp. Zellforsch. Jena* **15**, 373–380.
Fordham, M. G. C. (1925). *Aphrodite aculeata.* L.M.B.C. *Memoir* 27, *Proc. Liverpool Biol. Soc.* **40**, 121–126.
Fretter, V. (1953). *J. mar. biol. Ass. U.K.* **32**, 367–384.
Gamble, F. W. and Ashworth, J. H. (1898). *Quart. Jl. Microsc. Sci.* **41**, 1–42.
Gamble, F. W. and Ashworth, J. H. (1900). *Quart. Jl. Microsc. Sci.* **43**, 419–569.
Goodrich, E. S. (1893). *Quart. Jl. Microsc. Sci.* **34**, 387–402.
Goodrich, E. S. (1897). *Quart. Jl. Microsc. Sci.* **40**, 185–195.
Goodrich, E. S. (1898). *Quart. Jl. Microsc. Sci.* **41**, 439–457.
Goodrich, E. S. (1900). *Quart. Jl. Microsc. Sci.* **43**, 699–748.
Goodrich, E. S. (1904). *Quart. Jl. Microsc. Sci.* **48**, 233–245.
Goodrich, E. S. (1919). *Quart. Jl. Microsc. Sci.* **64**, 19–26.
Hanson, J. (1949). *Biol. Rev.* **24**, 127–173.

Hanson, J. (1951). *Quart. Jl. Microsc. Sci.* **92**, 255–274.

Hentschel, C. C. (1930). *Parasitology* **22**, 505–509.

Herlant-Meewis, H. and Nokin, A. (1963). *Ann. Soc. Roy. Zool. Belg.* **93**, 137–154.

Herlant-Meewis, H. and van Damme, N. (1962). *Mem. Soc. Endocr. Bristol* **12**, 287–295.

Herpin, R. (1921). *C. r. hebd. Séanc. Acad. Sci. Paris* **173**, 249–252.

Herpin, R. (1925). *C. r. hebd. Séanc. Acad. Sci. Paris* **180**, 864–866.

Hoffman, R. J., Mangum, C. P. and Black, R. E. (1968). *Amer. Zool.* **8**, 772.

Jourdan, E. (1887). *Bibl. l'école des hautes études, Sci. nat.* **34**, 1–66.

Kennedy, G. Y. and Dales, R. P. (1958). *J. Mar. Biol. Ass. U.K.* **37**, 15–31.

Kermack, D. (1955). *Proc. zool. Soc. Lond.* **125**, 347–381.

King, P. E., Bailey, J. H. and Babbage, P. C. (1969). *J. Mar. Biol. Ass. U.K.* **49**, 141–150.

Koechlin, N. (1966). *C. r. hebd. Séanc. Acad. Sci. Paris* **262D**, 1266–1269.

Kollmann, M. (1908). *Bull. Soc. Zool. Fr.* **34**, 149–155.

Kowalewsky, A. (1889). *Biol. Zbl. Leipzig* **9**, 43–47.

Kowalewsky, A. (1895). *C. r. 3rd Congr. Internat. Zool.* (1895), 526–530.

Kowalewsky, A. (1896). *Bull. Acad. Sci. St. Petersb.* **3**, 127–128.

Kryvi, H. (1972). *Sarsia* **49**, 59–64.

Kunstler, J. and Gruvel, A. (1898). *Archs. Anat. microsc.* **2**, 305–354.

Labbé, A. and Racovitza, E. G. *Bull. Soc. zool. Fr.* **22**, 92–97.

Lankester, E. R. (1873). *Proc. Roy. Soc.* **21**, 70.

Lankester, E. R. (1878). *Quart. Jl. Microsc. Sci.* **18**, 68–73.

Lee, E. (1912). *Jena Zeits. Naturw.* **48**, 433–478.

Liebman, E. (1946). *Growth* **10**, 291–330.

Lindroth, A. (1938). *Zool. Bidr. Uppsala* **17**, 367–497.

Liwanow, N. (1914). *Trudy Kazan Obschestest.* 46, 1–286.

McIntosh, W. C. (1907). *Ann. Mag. nat. Hist.* (7), **20**, 175–184.

Malaquin, A. (1893). *Mem. Soc. Sci. Arts, Lille* 1–477.

Mangum, C. P. and Dales, R. P. (1965). *Comp. Biochem. Physiol.* **15**, 237–257.

Mangum, C. P., Woodin, B. R., Bonaventura, C., Sullivan, B. and Bonaventura, J. (1975). *Comp. Biochem. Physiol.* **51A**, 281–294.

Margolis, L. (1971). *J. Fish Res. Bd. Canada* **28**, 1385–1392.

Marsden, J. R. (1966). *Can. J. Zool.* **44**, 377–389.

Marsden, J. R. (1968a). *Can. J. Zool.* **46**, 615–618.

Marsden, J. R. (1968b). *Can. J. Zool.* **46**, 619–624.

Mattisson, A. (1975). *Zool. Revy* **37**, 33–56.

Metchnikoff, E. (1893). "Lectures on the Comparative Pathology of Inflammation". Kegan Paul, London.

Meyer, E. (1887). *Mitt. Zool. Stat. Neapel* 7, 22–27.

Needham, A. E. (1970). *Symp. Zool. Soc. Lond.* **27**, 19–44.

Nicoll, P. A. (1954). *Biol. Bull., Woods Hole* **106**, 69–82.

Ochi, O. (1969). *Mem. Ehime Univ.* (2,B), **6**, 23–91.

Picton, L. J. (1897). *Proc. Trans. Liverpool biol. Soc.* **12**, 136–146.

Picton, L. J. (1898). *Quart. Jl. Microsc. Sci.* **41**, 263–302.

Pilgrim, M. (1965). *J. Zool.* **147**, 30–37.

Pilgrim, M. (1966). *J. Zool.* **149**, 242–261.

Pocock, D. M. E., Marsden, J. R. and Hamilton, J. G. (1971). *Comp. Biochem. Physiol.* **39A**, 683–697.

Potswald, H. (1969). *J. Morph.* **128**, 241–260.

Potswald, H. (1972). *J. Morph.* **137**, 215–228.
Prenant, M. (1929). *Bull. Lab. Mar. Mus. Hist. Nat. St. Servan* **4**, 4–5.
Probst, G. (1929). *Pubbl. Stat. Zool. Napoli* **9**, 317–387.
Quatrefages, A. de (1865). *In* "Histoire Naturelle des Annéles Marins et d'eau Douce", pp. 34–36. Rôret, Paris.
Racovitza, E. G. (1894). *C. r. hebd. Séanc. Acad. Sci. Paris* **118**, 153–155.
Racovitza, E. G. (1895). *C. r. hebd. Séanc. Acad. Sci. Paris* **120**, 464–467.
Rietsch, M. (1882). *C. r. hebd. Séanc. Acad. Sci. Paris* **92**, 926–929.
Romieu, M. (1921a). *C. r. Congrès Assoc. Anat., Paris* **82**, 187–193.
Romieu, M. (1921b). *C. r. hebd. Séanc. Acad. Sci. Paris* **173**, 246–249.
Romieu, M. (1921c). *C. r. hebd. Séanc. Acad. Sci. Paris* **173**, 786–788.
Romieu, M. (1921d). *C. r. hebd. Soc. Biol. Paris* **85**, 894–896.
Romieu, M. (1923). *Arch. Morph. exp. gén.* **17**, 1–336.
Rosa, D. (1896). *Mem. R. Acc. Sci., Torino* **46**, 149–177.
Schaeppi, T. (1894). *Jena Z. Naturw.* **28**, 249–293.
Schiller, I. (1907). *Jena Zeits. Naturw.* **43**, 293–320.
Schneider, G. (1899). *Z. wiss. Zool.* **66**, 497–520.
Schröder, G. (1886). *Zool. Inst. Wien, Rathenow Carl Koppel* 1–41.
Schroeder, P. C. (1967). *Amer. Zool.* **7**, 724.
Seamonds, B. and Schumacher, H. R. (1972). *Cytologia* **37**, 359–63.
Shafie, S. M., Vinogradov, S. N., Larson, L. and McCormick, J. J. (1976). *Comp. Biochem. Physiol.* **53B**, 85–88.
Sichel, G. (1961). *Boll. Zool.* **28**, 573–578.
Sichel, G. (1962a). *Boll. Accad. Gioenia Sci. nat.* (4) **6**, 321–323.
Sichel, G. (1962b). *Boll. Accad. Gioenia Sci. nat.* (4) **6**, 325–328.
Sichel, G. (1964). *Atti. Accad. Gioenia Sci. nat.* **8**, 86–93.
Siedlecki, M. (1903). *Ann. Inst. Pasteur* **17**, 449–462.
Smith, R. I. (1950). *J. Morph.* **87**, 417–466.
Stewart, F. A. (1900). *Ann. Mag. Nat. Hist. Lond* (7) **5**, 161–164.
Stunkard, H. W. (1943). *Biol. Bull., Woods Hole* **85**, 227–237.
Terwilliger, R. C. (1974). *Comp. Biochem. Physiol.* **48A**, 745–755.
Terwilliger, R. C. and Koppenhoffer, T. L. (1973). *Comp. Biochem. Physiol.* **45B**, 557–566.
Terwilliger, R. C., Garlick, R. L. and Terwilliger, N. B. (1976). *Comp. Biochem. Physiol.* **54B**, 149–153.
Thomas, J. A. (1930). *Archs Anat. microsc. Paris* **26**, 252–333.
Thomas, J. A. (1932). *Ann. Inst. Pasteur* **49**, 234–274.
Thomas, J. G. (1940). *Proc. Liverpool Biol. Soc.* **33**, 1–88.
Thouveny, Y. (1967). *Arch. Zool. exp. gén.* **108**, 347–386.
Tsusaki, T. (1928). *Acta. med. Keijo* **11**, 193–206.
Vejdovsky, F. (1882). *Denkschr. Acad. Wiss. Wien* **43**, 1–58.
Weber, R. E. (1978). *In* "Physiology of Annelids" (P. J. Mill, Ed.), pp. 393–446. Academic Press, London.
Weber, R. E. and Heidemann, W. (1977). *Comp. Biochem. Physiol.* **57A**, 151–155.
Wells, R. M. G. and Dales, R. P. (1975). *J. Mar. biol. Ass. U.K.* **55**, 211–220.
Wells, R. M. G. and Warren, L. M. (1975). *Comp. Biochem. Physiol.* **51A**, 737–740.
Williams, T. (1852). *Phil. Trans. Roy. Soc.* B. **142**, 595–653.
Zurcher, L. (1909). *Jena Zeits. Naturw.* **45**, 181–220.

4. Oligochaetes

E. L. COOPER AND E. A. STEIN

Department of Anatomy, School of Medicine, University of California, Los Angeles, California 90024, U.S.A.

CONTENTS

I. Introduction

The oligochaetes, as members of the phylum Annelida, are segmented protostomes, with a well-developed blood-vascular system and separate coelom. Although not as large a class as the polychaetes (approx. 3000 species, compared to 5300 in the polychaetes) (Dales, 1967), oligochaetes contain the most familiar of all annelids, the earthworms. Oligochaetes occupy a wide variety of habitats, ranging from soil to fresh and salt water. They also vary enormously in size, from less than 0·5 mm in members of the family Aeolosomatidae, to over 3 m for the giant earthworm of Australia (Barnes, 1974). The segmented coelom contains a fluid which not only acts as an hydraulic skeleton, but usually contains a number of free cells, or coelomocytes.

Since the blood vascular system is separate from the coelom, and may contain a distinct cell population, the two groups of cells will be treated separately. Relatively little is known of the blood cells, as compared with the coelomocytes, and of the latter, detailed information on structure and function is available for only a few families, primarily the Lumbricidae, Megascolecidae and Enchytraeidae. Coelomocytes of several species of Lumbricidae have recently been studied extensively in regard to their role(s) in cell-mediated immunity (Valembois, 1971a,b, 1974; Hostetter and Cooper, 1974; Lemmi *et al.*, 1974; Cooper, 1976a,b,c) and will be reviewed here.

Two major early works on the oligochaetes, Stephenson's "The Oligochaeta" (1930) and Avel's "Classe Des Annelides Oligochaetes" from "Traite De Zoologie" (1959), although somewhat outdated, contains comprehensive sections on blood cells and coelomocytes, and are excellent reference sources.

II. Structure of the circulatory system

A. The vascular system

Oligochaetes possess a well-developed, closed vascular system (Fig. 1). The basic plan consists of a contractile dorsal vessel, carrying blood anteriorly, which lies directly over and closely adherent to the alimentary tract. A ventral vessel, just under the intestine, carries blood posteriorly. Dorsal and ventral vessels are connected laterally, either directly by segmentally arranged pairs of commissural vessels, or indirectly by an intestinal plexus. These vessels carry blood to the various segmental organs and to the integument, receiving blood from the ventral vessel and draining into the dorsal vessel. Strongly contractile anterior commissural vessels, varying in number from one pair in *Tubifex* and most naidids to five pairs in *Lumbricus*, are referred to as "hearts" and function as accessory organs to the dorsal vessel in the movement of blood. Both the dorsal vessel and the hearts contain valves consisting of endothelial folds; in the dorsal vessel these are usually arranged segmentally. Various modifications of this plan include, in some species, an additional longitudinal vessel, the subneural vessel, which lies ventral to the nerve cord. Other species possess paired lateral neural vessels, which lie on either side of the nerve cord. A supra-intestinal vessel lying on the dorsal wall of the intestine may also be present in some worms, as well as lateral-oesophageal (Fig. 1) or extra-oesophageal vessels. An intestinal sinus is relatively common, the latter being situated between the epithelial and muscular layers of the intestinal wall (Stephenson, 1930; Dales, 1967). The histology of oligochaete blood vessels has been reviewed by Hanson (1949). Some variation occurs between species and within individuals, depending on the size and location of the vessel. The basic structure consists of a homogeneous connective-tissue layer, surrounded externally by a peritoneum of variable morphology and partially covered internally by a discontinuous endothelium. According to Hanson

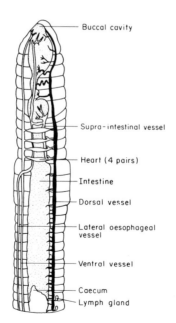

Fig. 1. Diagram of the circulatory system of *Pheretima hupeiensis* (modified from Grant, 1955).

(1949), endothelial cells have frequently been identified incorrectly as sessile blood leucocytes. In capillaries, the endothelium is apparently absent and the capillary wall consists of a basal lamina surrounded by perithelial cells (Hama, 1960; Chapron, 1970; Valembois, 1970).

B. The coelom

The coelom of oligochaetes is a fluid-filled cavity, lined with squamous epithelium, which lies between the intestine and the muscular body wall, and which runs the complete length of the body. At each segment, the cavity is subdivided by a transverse septum into chambers or compartments, each of which is connected to the exterior by a sphinctered dorsal pore. The transverse septa, covered with epithelium and containing complexly arranged muscle fibres, are perforated by one or more sphinctered apertures which allow the coelomic fluid to flow from one segment to another. These apertures most commonly occur around the passageway traversed by the ventral blood vessel and nerve cord (Edwards and Lofty, 1977).

III. Origin and formation of coelomocytes and blood cells

A. Coelomocytes

Coelomocytes are considered by most investigators to be derived from the epithelial lining of the coelomic cavity or from specialized structures associated with the epithelium (Kindred, 1929; Cameron, 1932; Liebmann, 1942; Duprat and Bouc-Lassalle, 1967; Valembois, 1971b). Embryologically, the coelom of oligochaetes follows the general plan of annelids, and is formed from the dorsal and ventral fusion or partial fusion of paired coelomic sacs which have their origins as bands of mesoderm. When the coelomic cavity develops, it lies within the mesoderm so that both the body wall and the endodermally derived digestive tube are lined with mesoderm. The resulting coelomic cavity is completely surrounded by mesoderm and the epithelium which lines the cavity is thought to be derived from it also (Meglitsch, 1967).

In many species of the genus *Pheretima*, and in *Megascolides australis*, *Dendrobaena rubida* and *Tubifex tubifex*, the coelomic epithelium has been modified to form specialized "lymphoid organs" considered by some investigators to be leucopoietic (Spencer, 1888; Beddard, 1890a,b; Thapar,

1918; Kindred, 1929). They are segmentally arranged, paired nodules, located throughout the intestinal region on either side of the dorsal vessel, and attached to the anterior face of the septa. They contain most of the different coelomocyte types as well as cells with a lymphocyte-like morphology, which were considered by Kindred (1929) to be stem cells. Although Kindred (1929) and Thaper (1918) believed the nodules to be leucopoietic, Schneider (1896) and Boveri-Boner (1920) regarded them primarily as phagocytic organs; there were no mitotic figures in any of the cells contained within the nodules.

In most species, however, coelomocytes are apparently derived directly from the coelomic epithelium, since it proliferates in response to the introduction of foreign substances, resulting in increased coelomocyte numbers (Cameron, 1932). Coelomocytes of Lumbricidae (the most extensively studied group) consists of two primary cell lines, amoebocytes and eleocytes. Amoebocytes are derived from the parietal and septal epithelia, and eleocytes from the epithelium covering the viscera and associated blood vessels (Cameron, 1932; Duprat and Bouc-Lassalle, 1967).

1. Amoebocyte line

Burke (1974b) postulated only one basic amoebocyte type, and that variations in coelomocyte morphology represent different developmental stages or states of activity of the same cell type. She described a single stem cell type, characterized by a particular type of cytoplasmic granule with an electron-transparent centre and electron-dense periphery. Other investigators have also identified only one stem cell type (Cuénot, 1898; Stephenson, 1930). Cameron (1932) reported that the proliferating epithelial tissue consists of a single cell type, similar in appearance to basophilic coelomocytes (hyaline amoebocytes, see below) which differentiates into the other cell types immediately after leaving the site of origin. The epithelial stem cell, as described by Chapron (1970), contains relatively little rough endoplasmic reticulum, but many free ribosomes and a number of dense osmiophilic granules.

Valembois (1971b) has outlined a more elaborate scheme of differentiation in *Eisenia foetida*, including three different stem cell types varying in ultrastructure and location. The first type, from the parietal peritoneum, has been termed "granuloblast" because of the presence of many small (~ 0.1 μm) osmiophilic granules. Granuloblasts are believed to be stem cells for the free coelomocytes termed "granulocytes" and "fibrillar leucocytes" (see section on Classification and structure of coelomocytes and blood cells). The second type occurs in the epithelial covering of the septa and of the larger vessels of the body wall. Called "hyaloblasts", they are relatively free of cytoplasmic granules and are stem cells for the "hyalocytes". A third, the "undifferentiated splanchnopleural cell", occurs at the junction of the peritoneal epithelium with the septa, and is considered the stem cell for "lymphocytoid cells", "splanchnopleural macrophages", "plasmacytoid macrophages", "vacuolar leucocytes" and "haemoglobin-producing cells".

2. Eleocyte line

The visceral peritoneum forms a highly specialized type of tissue, the chloragogen tissue, which surrounds not only the intestine, but also most of the dorsal blood vessel and the smaller vessels leading from the intestine to the dorsal vessel (Stephenson, 1930). When mature, the cells may become detached and enter the coelomic fluid, at which time they are frequently termed eleocytes (Liebmann, 1942; Duprat and Bouc-Lassalle, 1967; Chapron, 1970; Stang-Voss, 1971; Linthicum *et al.*, 1977a). During development, the endoplasmic reticulum is reduced, lipid-containing inclusions increase, and chloragosomes, haemoglobin-like crystals, and glycogen appear. Nearing its release into the coelomic fluid, little endoplasmic reticulum remains, and the cell is almost filled with chloragosomes and other types of inclusions (Chapron, 1970; Valembois, 1971b).

B. Blood cells

There is no definitive evidence for the origin of the cells found within the blood vessels. Some species, especially in the *Enchytraeidae*, *Tubificidae* and *Naididae*, are reported to possess few or no blood cells (Vejdovsky, 1884; Freudwiler, 1905). Vejdovsky (1884, 1905, 1906), Lison (1928), Stephenson (1930) and Hanson (1949) proposed that blood cells originate from cells of the vessel walls. DeBock (1900) found a special lymphoid organ in the mid-ventral line of the alimentary blood sinus of *Lumbriculus*, from which he believed blood cells proliferated. In several species of *Pheretima*, *Maoridrilus rosae*, *Pontodrilus michaelseni* and *Sparganopilus benhami*, specialized structures referred to as "blood glands" have been described by Beddard (1890a), Eisen (1895) and Stephenson (1924), and are believed by Stephenson to produce both blood cells and haemoglobin. Alternatively, Cuénot (1898) and Valembois (1970, 1971b) have suggested that blood cells are coelomocytes which have migrated into blood vessels by diapedesis. As supporting evidence, Cameron (1932) found after injecting India ink into the coelomic cavity, that cells containing ink particles were soon thereafter found within the blood vessels.

IV. Classification and structure of coelomocytes and blood cells

A. Introduction

The morphology of oligochaete coelomocytes has been investigated extensively in only five or six species, mostly from the Lumbricidae, Megascolecidae and Enchytraeidae (Tables I and II). Other families, particularly the aquatic groups, have received only superficial attention. Coelomocytes of *Lumbricus terrestris* and *Eisenia foetida* (both Lumbricidae), have been studied in greatest detail, and the major cell types found in these two species appear to be common to many members of other families whenever they have been examined in detail. Exceptions to this occur in some species (e.g. some members of Tubificidae and Naididae), which may have very few or no free coelomocytes, or which appear to have only one or two coelomocyte types (Sperber, 1948; Cook, 1969).

Morphological information on blood cells is scarce in the literature, and even when given for a particular species, has rarely included detailed descriptions. For this reason, we have included no classification system for blood cells; any such attempt must await further studies, and particularly a clarification of the relationship between coelomocytes and blood cells.

B. Coelomocyte classification-Lumbricidae

In the various studies of lumbricid coelomocytes, a number of different classification systems and terminologies have been used (Table I). They may be divided into two major categories, *amoebocytes* and *eleocytes*, with amoebocytes further divisible into *hyaline* and *granular* types. *Hyaline amoebocytes* are of more than one type, the two most commonly occurring in *L. terrestris* being designated as *basophils* and *neutrophils* (Stein et al., 1977), or, based on ultrastructural criteria, *lymphocytic coelomocytes* (Types I and II) and Type I *granulocytes* (Linthicum et al., 1977a), respectively. In *E. foetida*, hyaline amoebocytes, similar in morphology to basophils and neutrophils, have been termed *basophilic amoebocytes* and *macrophages*, respectively (Chapron, 1970), or *transportform amoebocytes*, *petaloid-form amoebocytes* and *stem cells*, the latter being small cells thought to be the precursors of either or both of the two other types (Stang-Voss, 1971).

Granular amoebocytes are also present in all lumbricid species. In such cells, granules are the dominant characteristic and nearly or completely fill the cells. Usually the granules are acidophilic, but in some cases are basophilic or of an intermediate staining reaction (Cameron, 1932; Stein et al., 1977). In *L. terrestris*, two types of acidophilic granular amoebocytes, or

TABLE I. Oligochaete coelomocyte classification-family Lumbricidae.

Species	Amoebocytes		Eleocytes	Miscellaneous	Reference
	Hyaline	Granular[a]			
Various Lumbricidae (review)	amoebocyte	vacuolar lymphocyte	eleocyte	mucocyte[d]	Rosa (1896)[b]
Lumbricus terrestris	small non-granular cell	small granular cell	chloragogen cell	spindle-shaped cell	Keng (1895)[b]
	large hyaline cell	large granular cell	eleocyte		Cuénot (1898)[b]
	amoebocytes (4 developmental stages)	amoebocyte with acidophilic granules	chloragogen cell		Kollmann (1908)[b]
	stage I[e] and II hyaline leucocytes	young granular leucocyte[e] adult granular leucocyte			
	basophilic cell	acidophilic cell	chloragogen cell		Cameron (1932)[b]
	stage I leucocyte[e]	eosinophilic leucocyte	eleocytes (5 stages)		Liebmann (1942)[b]
	amoebocyte petaloid amoebocyte		eleocyte		Vostal (1971)[b]
	procoelomocyte[e] intermediate cell	eosinophilic coelomocyte basophilic granular coelomocyte			
	small,[e] medium and large basophils	small[e] and large acidophilic cells transitional cell	chloragogen cell		Hostetter and Cooper (1974)[b]
	amoebocyte basophil neutrophil	amoebocyte acidophil (2 types) granulocyte	eleocyte chloragogen cell (2 types)		Burke (1974a)[c] Stein et al. (1977)[b]
	Type I lymphocytic coelomocyte[e] Type II lymphocytic coelomocyte Type I granulocyte	Type II granulocyte inclusion-containing cell	eleocyte		Linthicum et al. (1977a)[c]
Lumbricus rubellus, Lumbricus castaneus	amoebocytes (4 developmental stages)	amoebocyte with acidophilic granules	eleocyte		Cuénot (1898)[b]
	stage I[e] and II hyaline leucocytes	young granular leucocyte adult granular leucocyte	chloragogen cell		Kollmann (1908)[b]
	procoelomocyte[e] intermediate cell	eosinophilic coelomocyte basophilic granular coelomocyte	eleocyte		Vostal (1971)[b]

Species					Reference
Eisenia foetida	amoebocytes (4 developmental stages) stage I leucocyte[e] amoebocyte petaloid amoebocyte hyaline leucocyte	amoebocyte with acidophilic granules eosinophilic leucocyte	eleocyte chloragogen cell eleocytes (5 stages)		Cuénot (1898)[b] Liebmann (1942)[b] Duprat and Bouc-Lassalle (1976)[c] Chapron (1970)[c] Stang-Voss (1971, 1974)[c]
	basophilic amoebocyte macrophage stem cell[e] transport-form amoebocyte petaloid amoebocyte	granular leucocyte granular leucocyte vacuolar leucocyte	eleocyte eleocyte eleocyte		Vostal (1971)[b]
	procoelomocyte[e] intermediate cell	eosinophilic coelomocyte basophilic granular coelomocyte	eleocyte		
	hyalocyte	vacuolar leucocyte	chloragogen cell[c] eleocyte	haemoglobin-producing cell glycogen-containing cell	Valembois (1971b)[c]
	granulocyte[e]				
	fibrillar leucocyte lymphocytoid cell splanchopleural macrophage plasmocytoid macrophage				
Allolobophora terrestris, *Allolobophora caliginosa*, *Allolobophora chlorotica*, *Allolobophora icterica*, *Allolobophora venata*, *Allolobophora rosea*	amoebocytes (4 developmental stages)	amoebocyte with acidophilic granules	eleocyte chloragogen cell	mucocyte[d]	Cuénot (1898)[b]
Allolobophora longa *Allolobophora foetida*, *Dendrobaena platyura*	basophilic cell procoelomocyte[e] intermediate cell	acidophilic cell eosinophilic coelomocyte basophilic granular coelomocyte	chloragogen cell eleocyte		Cameron (1932)[b] Vostal (1971)[b]

[a] Predominant characteristic; [b] by light microscopy; [c] by electron microscopy; [d] in *Allolobophora rosea* only; [e] stem or precursor cell.

acidophils, have been described (Keng, 1895; Stein *et al.*, 1977) equivalent to *inclusion-containing cells* by electron microscopy (Linthicum *et al.*, 1977a), and similar to the *granular* or *vacuolar leucocytes* in *E. foetida* (Chapron, 1970). In addition, we (Stein *et al.*, 1977) have found a third type of granular amoebocyte, the *granulocyte*, which may be a developmental stage of one of the acidophils. The relationship between hyaline and granular amoebocytes has never been experimentally determined. However, a number of investigators have suggested that granular amoebocytes may be derived from small or medium-sized hyaline amoebocytes, based on the presence of a complete range of intermediate stages between the two types (Cuénot, 1898; Kollmann, 1908; Kindred, 1929; Cameron, 1932; Hostetter and Cooper, 1974; Stein *et al.*, 1977).

Eleocytes, the second major coelomocyte category, like the amoebocytes are found in all lumbricid worms. They are highly granular, but differ from the granular amoebocytes in morphology and granule characteristics. Eleocytes of different genera vary. In *E. foetida*, according to Liebmann (1942), eleocytes are long-lived cells undergoing a succession of changes after being released from the chloragogen tissue into the coelomic fluid. Such changes include expulsion of granules and biochemical alteration of the cytoplasmic inclusions (Liebmann, 1942; Valembois, 1971b). In *L. terrestris*, two types of eleocytes are present. The first (Type I chloragogen cell) is undoubtedly derived from the chloragogen tissue. Unlike the eleocytes of *E. foetida*, it appears to be a terminal cell degenerating soon after being released into the coelomic fluid (Cuénot, 1898; Stein *et al.*, 1977). The second type, designated Type II chloragogen cell (Stein *et al.*, 1977), is less obviously related to the chloragogen tissue and usually appears, in the coelomic fluid, to be an intact and non-degenerating cell.

In the lumbricid worm, *Allolobophora rosea* (but in no other species of this family), Rosa (1896) and Cuénot (1898) found a third major cell type, morphologically distinct from amoebocytes and eleocytes and referred to as a *mucocyte*. Mucocytes are large cells ($\sim$100 μm), non-amoeboid, lenticular in shape, and lacking visible inclusions except for a cluster of refringent globules surrounding the nucleus. The nucleus is large, central, oval or round, with one or two visible nucleoli. When coelomic fluid is discharged through the dorsal pores, the mucocyte changes shape to that of an elongated bi- or tri-polar cell, terminating in long, unbranched filaments.

C. Coelomocyte structure and cytochemistry-Lumbricidae

The coelomocytes of *L. terrestris* appear to be quite similar to those of other Lumbricidae (except for *A. rosea*, see above) (Table I). We have

TABLE II. Oligochaete coelomocyte classification—all other families.

| Species | Amoebocytes | | Eleocytes | Lamprocytes | Linocytes | Miscellaneous | Reference |
	Hyaline[a]	Granular[a]					
Ocnerodrilus lacuum	amoebocyte	eosinophil			nematocyte	mucocyte[k]	Eisen (1900)
Dichogaster crawi[b]	lymphocyte					mucocyte[k] morocyte	Eisen (1900)
Octochaetus multiporus[b]	amoebocyte	amoebocyte	eleocyte	lamprocyte	linocyte	spindle-shaped cell	Benham (1901)
Pheretima posthuma[b]	small nongranular cell	granular leucocyte	chloragogen cell	cell with refractile granules			Thapar (1918)
Pheretima heterochaeta, Pheretima hawayana Pheretima indica[b]	lymphocyte monocyte	granulocyte	chloragogen cell	lamprocyte	linocyte	detached peritoneal cell	Kindred (1929)
Pheretima sieboldi[b]	lymphocyte monocyte	granulocyte	chloragen cell	lamprocyte	linocyte	detached peritoneal cell	Ohuye (1937)
Enchytraeus hortensis[c, h]	amoeboid corpuscle	amoeboid corpuscle		oval corpuscle	thread corpuscle linocyte		Goodrich (1897)
Enchytraeus albidus[c]	amoebocyte		chloragogen cell	lamprocyte	linocyte		Fänge (1950)
Enchytraeus albidus[c]	procoelomocyte intermediate cell	eosinophilic coelomocyte granular coelomocyte	eleocyte	eleocyte	linocyte		Vostal (1971)
Enchytraeus fragmentosus[c]	amoebocyte		chloragogen cell	mucocyte		detached peritoneal cell	Hess (1970)
Drawida hattamimizu[d]	lymphocyte monocyte	basophilic granulocyte eosinophilic granulocyte	chloragogen cell	lamprocyte	linocyte	detached peritoneal cell	Ohuye (1934)
Pontoscolex corethrurus[e] Bothrioneurum iris[f] Nais sp.,[g] Slavinia sp.,[g]	amoebocyte	eosinophil		unnamed[j]		microcyte[i]	Eisen (1900) Aiyer (1925) Stephenson (1930)
Naidium breviseta[g] Pristina longiseta[g]	colourless corpuscle		granular corpuscle				

[a] Predominant characteristic; [b] Megascolecidae; [c] Enchytraeidae; [d] Moniligastridae; [e] Glossoscolecidae; [f] Tubificidae; [g] Naididae; [h] probably Enchytraeus albidus (see Stephenson, 1930); [i] non-nucleated corpuscles; [j] only one cell type briefly described; [k] possibly lamprocytes.

TABLE III. Coelomocyte characteristics—*Lumbricus terrestris*.[a]

Cell type	Frequency[b]	Cell size in µm	Nucleus[c]	Colour of cytoplasm	Granules[d]
Basophil	63·5 ± 6·1	5–30	4–8 central or eccentric dark blue-violet	blue	0·5, few dark blue
Neutrophil	18·0 ± 5·9	12–15	8–10 central or eccentric rose	pale blue, pink or lavender	0·5–1, few to moderate light blue, pink or lavender
Acidophil Type I	6·2 ± 3·4	10–30	5–9 usually eccentric dark red-violet	blue	1–2, many pink to red
Acidophil Type II	0·7 ± 0·5	10–15	6–8 central or eccentric dark red-violet	blue	2–4, moderate to many pink to red
Granulocyte	8·1 ± 6·6	8–40	5–9 often eccentric dark violet	blue	0·5–2, variable pink, lavender, blue
Chloragogen Type I	see Type II (below)	10–25 by 30–60	6–7 usually eccentric rose	pale blue	1–2, many medium blue
Chloragogen Type II	1·4 ± 0·9[e]	12–20	7–8 usually eccentric rose	blue	0·5–2, many dark blue-violet
Transitional	1·9 ± 1·0	variable	variable	variable	variable

[a] Cells stained with Wright's stain; [b] per cent of the total population, expressed as mean value ± S.D.; [c] size in µm, position in cell and colour; [d] approximate size in µm, number, colour; [e] combined frequency for chloragogen Types I and II. (From Stein *et al.*, 1977).

studied in detail the morphology of coelomocytes of *L. terrestris* both by light microscopy, using differentially stained and living preparations (Stein *et al.*, 1977 (Table III), and by electron microscopy (Linthicum *et al.*, 1977a). We have also employed cytochemical techniques to characterize each of the coelomocyte types in terms of its major biochemical components (Stein and Cooper, 1978).

Using a modified Wright's staining procedure (Stein *et al.*, 1977) we have classified the coelomocytes of *L. terrestris* into five major categories. These are the basophils and neutrophils (together = hyaline amoebocytes), acidophils and granulocytes (together = granular amoebocytes), and the chloragogen cells (eleocytes). Within each category we have then integrated, wherever possible, the ultrastructural and cytochemical properties of each cell type. The equivalent terms for the various cell types in the light and electron microscopes are given in Table IV.

1. Basophils

(*a*). *Light microscopy.* Basophils are the most numerous of the coelomic cells and vary in size from 5–30 μm (Fig. 2a). They strongly adhere to one another and frequently form sheets or clumps. The cytoplasm is moderately to strongly basophilic, often containing small (approximately 0·5 μm) dark-blue granules (Table III). Live cells sometimes contain a few small reddish-brown granules in addition to clear vacuoles of variable size

TABLE IV. Summary of various cell types present in *Lumbricus terrestris*.

General name	Light microscopy designation (Wrights stain)	Electron microscopy designation
Hyaline amoebocytes	1. basophils ⟶	lymphocytic coelomocytes of 2 types
	2. neutrophils ⟶	Type I granulocytes
Granular amoebocytes	3. acidophils (2 types) ⟶	inclusion-containing coelomocytes
	4. granulocytes – – – – – – – →	not identified
	5. Not identified – – – – – – →	Type II granulocytes
Eleocytes ⟶	6. chloragogen cells ⟶ (eleocytes) (2 types)	chloragogen cells (eleocytes) (1 type only)

(0·5–6 μm); some of the latter are probably phagocytic vacuoles, since they often contain bacteria or chloragogen granules. Medium-sized and large basophils produce both petaloid and long filamentous pseudopodia (Fig. 2b), while smaller basophils usually produce short spikes or no observable pseudopodia. Intermediate forms between basophils and non-granular acidophils, and between granular acidophils and granulocytes, are fairly common, but whether there is a developmental relationship, as proposed by Cameron (1932) and Hostetter and Cooper (1974), is still unclear.

(*b*). *Cytochemistry*. Basophils contain cytoplasmic protein, PAS-positive non-glycogen carbohydrates, and small lipid-positive granules (Table V). They also contain moderate to large amounts of cytoplasmic RNA, primarily responsible for their basophilia (Duprat and Bouc-Lassalle, 1967; Stein and Cooper, 1978). Neither cytoplasm nor granules contain acid mucopolysaccharides, haemoglobin, nor iron, properties which apply also to the cytoplasm and granules of all other coelomocyte types except chloragogen cells. Clear vesicles or vacuoles are of two types; the more numerous one is apparently filled with glycogen (Cuénot, 1898; Stang-Voss,

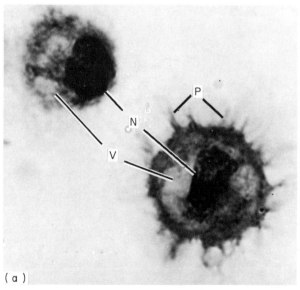

(a)

Fig. 2(a). Two intermediate-size basophils of *Lumbricus terrestris* containing vacuoles (V). The nucleus (N) in the smaller is eccentric but in the larger is central, and the latter has several small filamentous pseudopodia (P). Wright's stain. (Reproduced with permission from Stein *et al.*, 1977). × 2000.

1971; Stein and Cooper, 1978). The other group, probably phagocytic vacuoles, contains acid phosphatase and β-glucuronidase, two enzymes which are also localized within smaller (about 0·5 μm), discrete granules presumed to be lysosomes (Stang-Voss, 1971; Stein and Cooper, 1978). These granules also stain supravitally with neutral red and acridine orange, further indication that they are lysosomes (Stein, unpublished). The amount of acid phosphatase per cell tends to be higher in basophils than in other coelomocytes, and more basophils contain this enzyme than the general population (74 compared to 60%). No peroxidase has been found in basophils or the other coelomocyte types (Stein and Cooper, 1978), except for low levels detected in eleocytes by Fischer and Horvath (1978).

(c). *Electron microscopy (lymphocytic coelomocytes)*. Based on ultra-structural criteria, basophils have been termed *lymphocytic coelomocytes* and occur in two forms (Linthicum *et al.*, 1977a). Type I lymphocytic coelomocytes are usually spherical or ovoid and the scanty cytoplasm is essentially devoid of elaborate organelles, but mitochondria, usually small (0·15–0·20 μm) with longitudinal cristae, are moderate in numbers (Fig. 2c).

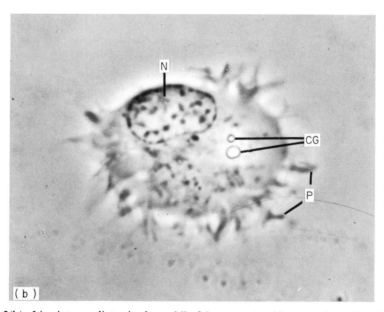

Fig. 2(b). Live intermediate-size basophil of *L. terrestris* with eccentric nucleus. The pseudopodia (P) are mixed filamentous and petaloid, and phagocytized chloragogen granules (CG) are present. Phase contrast. (Reproduced with permission from Stein *et al.*, 1977). × 2500.

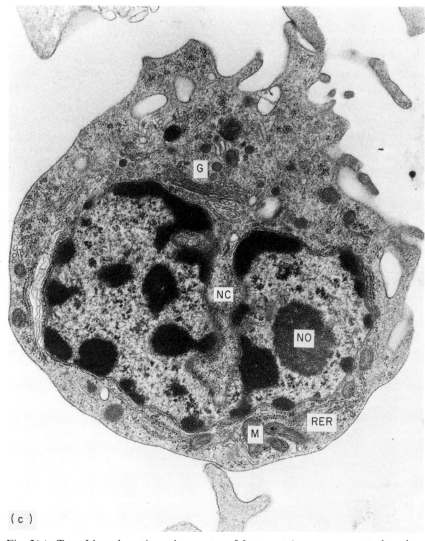

(c)

Fig. 2(c). Type I lymphocytic coelomocytes of *L. terrestris* possess a central nucleus with condensed nucleolus (NO) and a nuclear cleft (NC). Small amounts of rough endoplasmic reticulum (RER), few mitochondria (M), but a well-defined Golgi (G) complex characterize the cytoplasm. (Reproduced with permission from Linthicum *et al.*, 1977a). × 22 800.

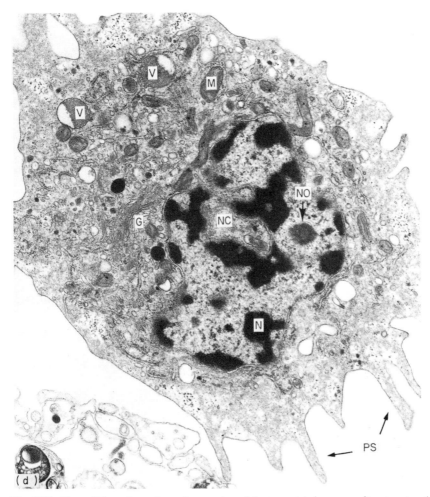

Fig. 2(d). Type II lymphocytic coelomocytes of *L. terrestris* have an ultrastructural appearance similar to Type I except for their slightly larger size and the presence of vacuoles containing dark flocculent material (V). Long, thin pseudopodia (PS) are prominent features. Nucleus (N), nucleolus (NO), nuclear cleft (NC). Golgi complex (G). (Reproduced with permission from Linthicum *et al.*, 1977a). × 13 200.

Type II lymphocytic coelomocytes tend to be slightly larger than Type I and contain many free and membrane-associated ribosomes (Fig. 2d). The central nucleus of Type II cells is essentially similar to that of Type I cells. Unlike the smaller Type I lymphocytic coelomocytes, pseudopodial formations on Type II cells are numerous, suggesting that they are mobile and phagocytic. Lymphocytic coelomocytes generally show certain structural similarites to immature lymphocytes of primitive vertebrates (Linthicum, 1975), and even mammals (Bessis, 1972). They are also similar to other earthworm coelomocytes (e.g. *E. foetida*), called *basophilic amoebocytes* by Chapron (1970), *young amoebocytes* or *stem cells* by Stang-Voss (1971), and *lymphocytoid cells* by Valembois (1971a,b) (Table I). Type II lymphocytic coelomocytes may represent a more mature form than the Type I, since they are slightly larger and contain more organelles.

TABLE V. Cytochemical reactions of coelomocytes of *Lumbricus terrestris* (from Stein and Cooper, 1978).

Cytochemical stain	Substance tested for		Basophils	
		cyto	small gran[a] (0–2)	gly vac (0–2)
Pyronin	presumptive ribonucleic acid	1–3	0	0
Pyronin after ribonuclease	confirmation of ribonucleic acid by its specific removal	0	0	0
Hg-bromphenol blue	proteins (general)	1–2	2	0
Biebrich Scarlet				
Methanol-fixed pH 8·0	basic proteins (staining at higher pH's and after	0–2	0	0
9·5	formalin fixation indicates presence of more	0–2	0	0
10·5	strongly-basic amino acids)	0–1	0	0
Formalin-fixed pH 9·5		0–1	0	0
PAS	most carbohydrates (except acid mucopolysaccharides)	1–2	1–2	3
PAS after amylase	glycogen	1–2	1–2	0
Iodine	glycogen	0	0	2
Alcian blue pH 1·0	sulphated acid mucopolysaccharides	0	0	0
2·5	most acid mucopolysaccharides	0	0	0
PAPS	acid mucopolysaccharides	0	0	0
Sudan black B	lipids and phospholipids	1–2	2–3	0
Sudan black B				
after ethanol	alcohol-soluble lipids	0	0	0
Leuco patent blue V	haemoglobin pseudoperoxidase	0	0	0
Prussian blue	ferric iron	0	0	0
Turnbull blue	ferrous iron	0	0	0
Simultaneous coupling-azo				
dye procedure	acid phosphatase	0	3	0
Simultaneous coupling-azo				
dye procedure	β-glucuronidase	0	3	0

Strength of reactions are rated on a 0–3 scale: 0, negative; 1, weakly positive; 2, moderately positive; 3, strongly positive; ±, uncertain reaction. Numbers directly underneath each heading for granules or other inclusions indicate the number of inclusions/cell: 0, none; 1, few; 2, moderate number; 3, many.

Abbreviations: cyto, cytoplasm; gly, glycogen; gran, granules; pred. predominant; vac, vacuoles.

2. Neutrophils

(a). *Light microscopy.* Highly variable in size, neutrophils may range from 12 μm to as large as 40 or 50 μm (Fig. 3a) (Table III). Cell margins are irregular and indistinct, and blunt pseudopodia are produced, which sometimes terminate in thin, finely drawn filaments of varying lengths (Fig. 3b). In live preparations the particulate endoplasm is surrounded by a clear zone of non-particulate ectoplasm. Small, dark granules are present in moderate numbers, along with many clear vesicles of approximately 1–2 μm, some of which contain phagocytized material.

(b). *Cytochemistry.* The cytoplasm of neutrophils contains small amounts of RNA, non-glycogen carbohydrate and protein, and little or no lipid (Table V). Small (ca. 0·5 μm) sudanophilic granules are often numerous and are similar in appearance and staining properties (although less strongly

| Neutrophils | | | Acidophils | | | | Granulocytes | | Chloragogen cells | | | |
| | | | Type I | | Type II | | | | Type I | | Type II | |
cyto	gly gran (0–2)	small gran[b] (0–3)	cyto	pred gran[c,d] (3)	cyto	pred gran[c,d] (3)	cyto	pred gran[d] (2–3)	cyto	gran (3)	cyto	gran (3)
1	0	0	1	1	1	±	1	0–1	0	0	0	3
0	0	0	0	2	0	±	0	1	0	0	0	3
1	0	1	1, 3	3	1, 3	2	1–2	2–3	1–2	1–2	1–2	0–2
0–1	0	0	0–1	2–3	0–1	2–3	0–1	2–3	0–2	0–1	0	0
0	0	0	0–1	2–3	0–1	1–2	0–1	2	0–2	0–1	0	0
0	0	0	0	2	0	1–2	0	1–2	0–2	0	0	0
0	0	0	0	2–3	0	1–2	0	1–2	0–1	0	0	0
1	2–3	1	1, 3	2–3	1, 3	1–2	1	2	0–3	0–3	0–2	0–3
1	0	1	1, 3	2–3	1, 3	1–2	1	2–3	0–3	0–3	0–2	0–3
0	1–2	0	0	0	0	0	0	0	0–2	0	0–2	0
0	0	0	0	0	0	0	0	0	0–2	0	0–2	0
0	0	0	0	0	0	0	0	0	0–3	0	0–3	0
0	0	0	0	0	0	0	0	0	0–3	0	0–3	0
0–1	0	1	0–1	0	0	0	0	0	0	3	0	3
0	0	0	0	0	0	0	0	0	0	0	0	3
0	0	0	0	0	0	0	0	0	0	2[e]	0	2[e]
0	0	0	0	0	0	0	0	0	0	2[e]	0	2[e]
0	0	0	0	0	0	0	0	0	0	2[e]	0	2[e]
0	0	3	0–1	0	0–1	0	0–1	0	0–1	0	0–2	0
0	0	3	0–1	0	0–1	0	0–1	0	0–1	0	0–2	0

[a] Does not include few 1μm granules, which stain only with Sudan Black B; [b] does not include Biebrich Scarlet-positive granules (see text); [c] does not include glycogen granules; for reactions, see column listed under neutrophils; [d] does not include small granules; for reactions, see column listed under basophils; [e] only small to moderate numbers of granules are positive.

stained) to the small, lipid-positive granules of basophils and other coelomocytes. Glycogen quantities vary and are often large. Acid phosphatase and β-glucuronidase are localized in discrete granules which vary widely in number but occasionally are quite numerous. The percentage of acid phosphatase-positive neutrophils is relatively high (64%) and is exceeded only by basophils.

(c). *Electron microscopy (Type I granulocytes).* Type I granulocytes are moderately large cells, 9–15 μm in diameter (Fig. 3c), and correspond to neutrophilic coelomocytes of light microscopy (Linthicum *et al.*, 1977a). The nucleus, usually in a central position, is large (6–7 μm) and contains primarily light-staining euchromatin. The endoplasm contains most of the organelles, including broad bundles of microfilaments. Small (0·1–0·3 μm) dark-staining granules (Fig. 3d), mostly ovoid but sometimes dumbbell-shaped, are numerous. The highly electron-dense granules resemble lysosomes or the specific and azurophilic granules of vertebrate neutrophils (Bainton and Farquhar, 1966). Dispersed in and around the endoplasm are glycogen packets containing both α and β particles, which may represent the end result of active phagocytosis, as occurs in clam granulocytes (Cheng and Cali, 1974). Type I granulocytes are similar to *macrophages* (Chapron, 1970; Valembois, 1971b) and *petaloid phagocytes* (Stang-Voss, 1971) of *E. foetida* (Table I).

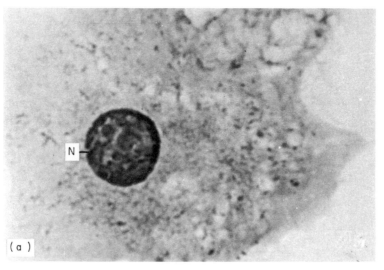

(a)

Fig. 3(a). Neutrophil of *L. terrestris* with a large nucleus (N) and lightly-staining cytoplasm. Wright's stain. (Reproduced with permission from Stein and Cooper, 1978). × 2000.

3. Acidophils

(a). *Light microscopy*. Acidophils are usually granular cells and, in *L. terrestris*, are of two types, varying in granule size and other structural properties (Table III). Size varies from 10 to 30 μm, with Type I (Fig. 4a) tending to be larger (15–30 μm) and Type II smaller (10–15 μm) (Fig. 4b). Occasionally, the cytoplasm of smaller cells, and more rarely of larger cells, may appear to be nearly homogeneous and without granulation, but is still acidophilic. Such cells are possibly precursors to the granular form of acidophils (Stein *et al.*, 1977). In live preparations, acidophils display both filamentous and petaloid pseudopodia, but these are usually smaller and less numerous than those found on basophils (Figs 4c,d). Acidophils of both types have been observed to release the granular material, first as small blebs on the cell surface (Fig. 4d) which then enlarge and finally pinch off. The function of the secreted material is unknown.

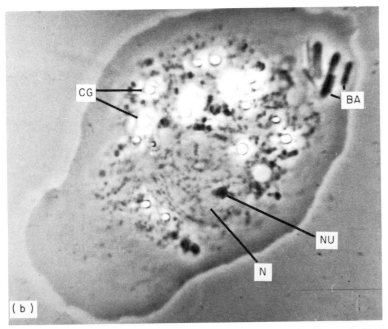

Fig. 3(b). Living neutrophil of *L. terrestris* containing phagocytized bacteria (BA) and chloragogen granules (CG). Large nucleus (N) contains a nucleolus (NU) and is relatively non-granular. Phase contrast. (Reproduced with permission from Stein *et al.*, 1977.) × 2000.

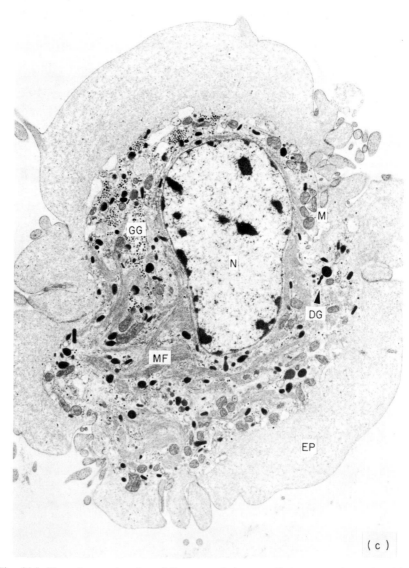

(c)

Fig. 3(c). Type I granulocytes of *L. terrestris* have a distinct ectoplasm devoid of organelles (EP), and possess blunt pseudopodia. The perinuclear endoplasm contains electron-dense granules (DG), glycogen granules (GG), mictochondria (M), and extensive microfilaments (MF). (Reproduced with permission from Linthicum *et al.*, 1977a). × 8200.

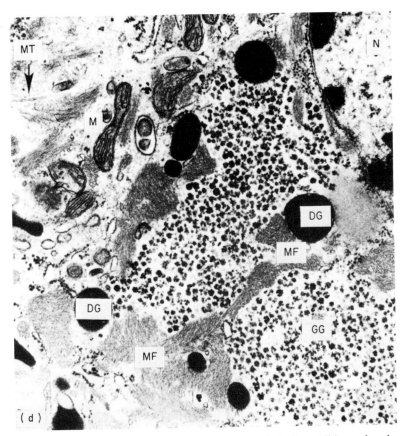

Fig. 3(d). Type I granulocytes of *L. terrestris* characteristically exhibit an abundance of electron-dense granules (DG). Bundles of microfilaments (MF) in the perinuclear endoplasm surround packets of cytoplasmic glycogen particles (GG). Mitochondria (M) and microtubules (MT) are also visible. Nucleus (N). (Reproduced with permission from Linthicum *et al.*, 1977a). ×35 100.

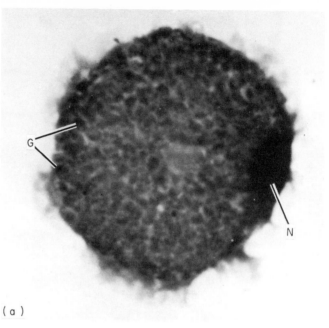

(a)

Fig. 4(a). Type I acidophil of *L. terrestris*, containing many granules of the predominant type (G) and a highly eccentric nucleus (N). Small pseudopodia project from the periphery of the cell. Wright's stain. (Reproduced with permission from Stein and Cooper, 1978). × 2200.

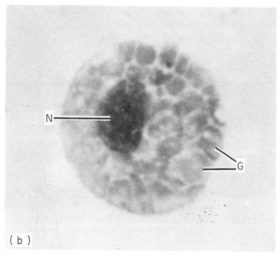

(b)

Fig. 4(b). Type II acidophil of *L. terrestris*, filled with predominant granules (G). Nucleus (N). Wright's stain. (Reproduced with permission from Stein and Cooper, 1978). × 3000.

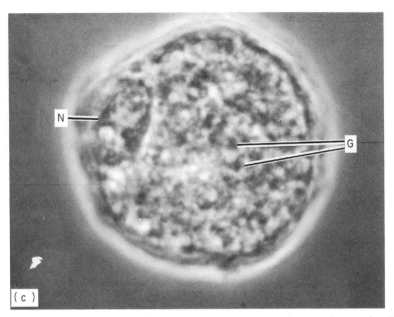

Fig. 4(c). Living Type I acidophil of *L. terrestris*, containing small granules (G), with the nucleus (N) at the extreme edge of the cell. Phase contrast. (Reproduced with permission from Stein *et al.*, 1977). × 2200.

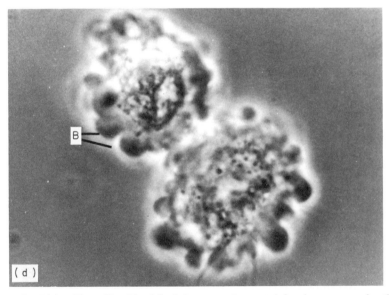

Fig. 4(d). Living Type II acidophil of *L. terrestris*, containing larger granules than Type I; some project as blebs (B) from cell surface. Phase contrast. (Reproduced with permission from Stein *et al.*, 1977). × 2500.

(*b*). *Cytochemistry*. Cytochemical reactions of the cytoplasm of acidophils are often obscured by the granules, but when observable are nearly identical to those of larger basophils. The predominant granules of both acidophil types react similarly with most staining procedures, except that granules of Type II usually stain more weakly than Type I. Both contain a neutral mucopolysaccharide or glycoprotein and a strongly basic protein, but no lipid, acid mucopolysaccharides, haemoglobin, nor iron (Table V). Granules of Type I acidophils also exhibit a slight pyroninophilia that is unaffected by ribonuclease extraction. Acidophils also possess a second much less numerous population of small granules similar to the small sudanophilic granules of

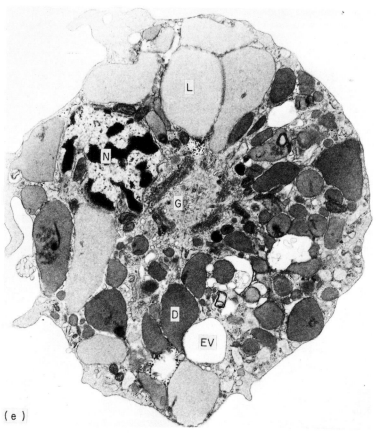

(e)

Fig. 4(e). Inclusion-containing coelomocyte (acidophil) of *L. terrestris* with a well-developed Golgi complex (G) and with numerous inclusion bodies which show varying degrees of dark (D) or light (L) staining. Empty vesicles (EV) are also visible. (Reproduced with permission from Linthicum *et al.*, 1977a). × 10 000.

basophils. A third small population of granules, probably glycogen, is more variable in size (0·5–1·5 μm) and irregular in outline. Acid phosphatase and β-glucuronidase are present in both acidophil types with low-to-moderate frequency, but never in large amounts. The enzymes usually appear to be distributed diffusely throughout the cytoplasm, although β-glucuronidase can occasionally be seen within quite small (<0·5 μm) granules.

(c). *Electron microscopy (inclusion-containing coelomocytes).* Coelomocytes that apparently synthesize, store and secrete large inclusions, i.e. acidophils, have been classified ultrastructurally as inclusion-containing coelomocytes. In *L. terrestris*, only one type has been identified by electron microscopy, and this may be equivalent to the Type II acidophil of light microscopy. They are similar to coelomocytes of *E. foetida* which have been termed *late-stage eleocytes* (Stang-Voss, 1971), *erythroid cells* (Stang-Voss, 1974) and *granular* or *vacuolar leucocytes* (Chapron, 1970) (Table I). These cells tend to be spherical and vary from 5 to 8 μm, possibly reflecting immature and mature forms. They contain many large, electron-dense inclusions of varying size and of fine granular appearance (Fig. 4e); there is little endoplasmic reticulum. The Golgi apparatus is well-developed, and apparently the inclusion bodies are packaged in the convex side of this organelle. Small Golgi vesicles contain an electron-dense material with distinct limiting membranes. When these vesicles reach the cytoplasmic periphery, they become lighter-staining and larger in size, so that putative mature cells have fewer electron-dense inclusions than the immature forms. Granulation does not appear to change except for decreased electron-opacity. Once at the cytoplasmic periphery larger inclusions may undergo expulsion from the cell.

4. Granulocytes

(a). *Light microscopy.* Granulocytes contain numerous easily ob-servable granules scattered throughout the light blue cytoplasm (Fig. 5a) (Table III) and in live cells numerous vacuoles are visible (Fig. 5b). Granules (0·5–2 μm) are most frequently acidophilic but may also be basophilic, or of an intermediate lavender colour. As in the case of acidophils, small surface blebs are occasionally present.

(b). *Cytochemistry.* The predominant (i.e. most numerous) granules of granulocytes usually react similarly to the predominant granules of Type I acidophils with most cytochemical procedures, except there appears to be less basic protein (Table V). Acid phosphatase occurs relatively infrequently in small amounts per cell. As in acidophils it appears to be in a non-granular form. Beta-glucuronidase is distributed similarly, but in slightly larger amounts and occasionally it is visible within 0·5 μm granules.

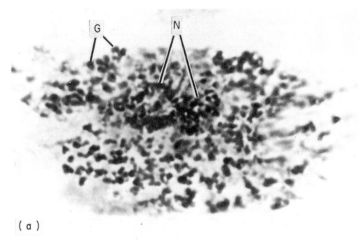

(a)

Fig. 5(a). Binucleate form of granulocyte of *L. terrestris*, which occurs occasionally in this cell type. Note amoeboid appearance and the scattering of granules (G). Nuclei (N). Wright's stain. (Reproduced with permission from Stein *et al.*, 1977). × 2000.

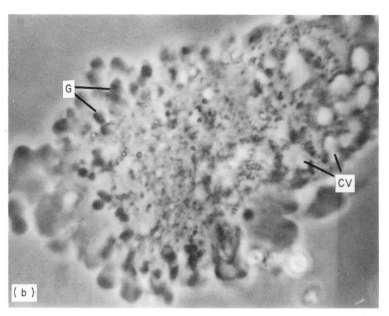

Fig. 5(b). Living granulocyte of *L. terrestris* containing numerous clear vesicles (CV) and granules of varying sizes (G). Phase contrast. (Reproduced with permission from Stein *et al.*, 1977). × 2500.

5. Type II granulocytes

Type II granulocytes have been identified only by electron microscopy (Linthicum et al., 1977a). They are stellate in appearance due to elaborate pseudopodial extensions (Fig. 6) and granules and empty vacuoles are distinguishing features. They are medium in size (7–9 μm in diameter for the body of the cell), but the extent of the pseudopodia can vary greatly. The nucleus is slightly eccentric and contains peripherally condensed chromatin; a nucleolus is occasionally visible and there are small amounts of rough endoplasmic reticulum and Golgi vesicles. A moderate number of mitochondria are visible, but there are no bundles of microfilaments. Two types of granules are present: (1) small (0·1–0·3 μm) membrane-limited, electron-dense, resembling the ovoid granules of Type I cells, but less abundant, and probably lysosomes; and (2) less opaque and somewhat larger (0·2–0·8 μm) granules containing a homogeneous material with a distinct limiting membrane. Both kinds of granules are found throughout the cytoplasm and in thin pseudopodial extensions. Peripheral portions of the cytoplasm sometimes contain small packets of glycogen particles. There are no well-defined endoplasmic and ectoplasmic regions.

Type I and Type II granulocytes are not, we believe, different forms of the same coelomocyte, but rather entirely different lineages as evidenced by their staining characteristics and fine structure. Type I granulocytes appear to be identical to neutrophils by light microscopy; however, we have been unable to definitely equate Type II granulocytes with a specific coelomocyte type by light microscopy. Perhaps they are equivalent to acidophilic granulocytes (see section on granulocytes) since both coelomocytes contain a granular cell product, numerous vacuoles, and a highly irregular cell outline.

6. Chloragogen cells (eleocytes)

(a) Light microscopy. Chloragogen cells, or eleocytes, occur in one or perhaps two forms depending upon various morphological differences (Table III). Both are highly granular cells with no pseudopodia visible by light microscopy. The first type is an oblong cell ranging in size from 10–25 μm wide by 30–60 μm long (Fig. 7a). They are frequently found in the coelomic fluid in a state of partial disintegration that apparently occurs naturally, since in freshly drawn coelomic fluid other coelomocytes often contain phagocytized chloragogen granules. The second type is smaller (12–20 μm), less numerous, and usually occurs in clusters of two or more cells. In living cells the granules of Type I are yellow-brown and highly refractile, while those of Type II are reddish-brown and less refractile. Cells

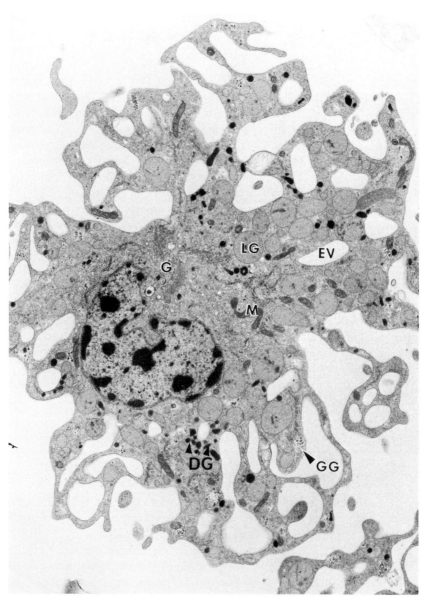

Fig. 6. Type II granulocytes of *L. terrestris* are stellate in appearance, due to their numerous pseudopodia, and empty vesicles (EV) appear in the peripheral cytoplasm. A Golgi apparatus (G) is present, as are small numbers of glycogen granules (GG). Both dense (DG) and light-staining granules (LG) are present. (Reproduced with permission from Linthicum *et al.*, 1977a). × 9000.

scraped from the visceral peritoneum, their source of origin (Fig. 7b), are usually Type I.

(b). *Cytochemistry*. Both chloragogen cell types show similar but not identical cytochemical reactions (Table V). A greater degree of variability often occurs among cells of one type than between cells of the two different types. The cytoplasm of both types contains variable amounts of protein and PAS-positive, non-glycogen carbohydrate. Glycogen is also present, often in large amounts (Cuénot, 1898; Liebmann, 1942; Semal Van-Gansen, 1956; Roots, 1960; Duprat and Bouc-Lassalle, 1967; Linthicum et al., 1977a; Stein and Cooper, 1978). Both types lack detectable cytoplasmic RNA, but in contrast to all other coelomocytes they frequently (but not always) show high concentrations of acid mucopolysaccharides.

Chloragogen vesicles or granules (chloragosomes) contain variable amounts of PAS-positive, non-glycogen carbohydrate, but no acid mucopolysaccharides. They also contain protein in variable amounts— approximately 16% by dry weight (Urich, 1960). Lipids constitute up to

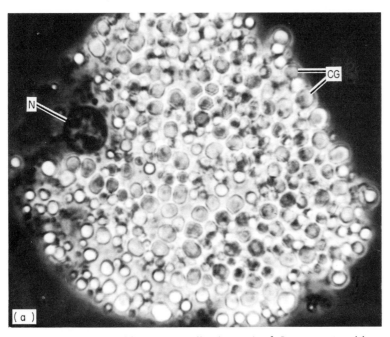

Fig. 7(a). Living Type I chloragogen cell (eleocyte) of *L. terrestris* with small, eccentric nucleus (N). Granules (CG) are refractile and fill the cell. Phase contrast. (Reproduced with permission from Stein et al., 1977). × 2200.

61 % of the dry weight of chloragosomes (Urich, 1960), and are apparently complex, consisting of phospholipids (Semal Van-Gansen, 1956; Urich, 1960; Roots and Johnston, 1966), carotenoids (Urich, 1960; Roots and Johnston, 1966; Fischer, 1975), a lipofuscin-like substance (Semal Van-Gansen, 1958; Fischer, 1977), flavins (Fischer, 1972; Moment, 1974), flavones, and flavenols (Roots and Johnston, 1966; Fischer, 1972). Various substances related to nitrogen metabolism have been reported, namely guanine (Willem and Minne, 1900), later refuted by Abdel-Fattah (1955), Semal Van-Gansen (1956) and Roots (1960); heteroxantine (Semal Van-Gansen, 1956), later refuted by Abdel-Fattah (1955) and Roots (1960); urea (Heidermanns, 1937; Cohen and Lewis 1949; Abdel-Fattah, 1955; Semal Van Gansen, 1956), and ammonia (Semal Van-Gansen, 1956). Granules of Type II (but not Type I) cells contain an unidentified strongly pyroninophilic, ribonuclease-stable substance.

A number of enzymes have been reported: an arginase-like enzyme (Abdel-Fattah, 1955), non-specific esterase, phosphorylase, and alkaline phosphatase (Semal Van-Gansen, 1956, 1958), acid phosphatase and β-glucuronidase (Semal Van-Gansen, 1956, 1958; Varute and More, 1972,

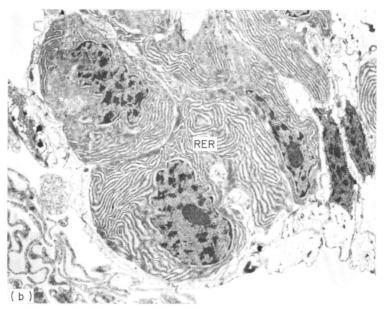

Fig. 7(b). Developing chloragogen cells within the chloragogen tissue of *Eisenia foetida*. Rough endoplasmic reticulum (RER) is highly developed. (Reproduced with permission from Chapron, 1970). × 4500.

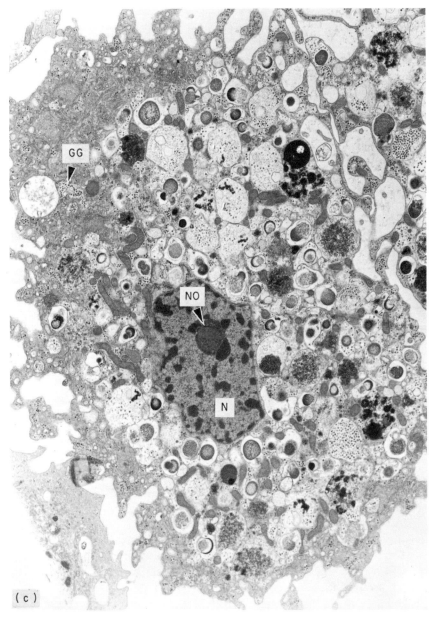

(c)

Fig. 7(c). Eleocytes of *L. terrestris* are characterized by their large size and the variety of vacuoles, inclusions and granules. Glycogen granules (GG) appear both in vesicles and free within the cytosol. Nucleus (N), nucleolus (NO). (Reproduced with permission from Linthicum *et al.*, 1977a). × 7600.

1973; Stein and Cooper, 1978); extra-peroxisomal catalase and peroxidase (Fischer and Horvath, 1978). We (Stein and Cooper, 1978) found no peroxidase nor did Prentø (1979), but we did find a non-enzymic pseudo-peroxidase, perhaps related to the metalloporphyrins reported by Fischer (1975) or to haemoglobin or ferritin, identified by its crystalline structure (Lindner, 1965; Chapron, 1970; Stang-Voss, 1971).

Chloragogen cells accumulate a number of metallic ions, including iron (Schneider, 1896; Matsumoto, 1960; Stein and Cooper, 1978; Prentø, 1979), lead (Ireland and Richards, 1977), calcium, zinc and magnesium (Prentø, 1979), and also silicate-containing compounds (Semal Van-Gansen, 1956; Roots, 1960).

(c). *Electron microscopy.* Eleocytes (chloragogen cells), as identified by electron microscopy, are large cells with a low nuclear:cytoplasmic ratio (Fig. 7c). Little or no rough endoplasmic reticulum or Golgi complexes are visible, but the cytoplasm is packed with a variety of inclusions, vesicles and granules, some of which contain what appears to be glycogen. Glycogen particles are also present in the cytoplasm. Other vacuoles contain electron-dense globules of many shapes and sizes. Eleocytes do not contain lysosome-like granules similar to those observed in Types I and II granulocytic coelomocytes. By electron microscopy, we were not able to distinguish a second type of eleocyte equivalent to the Type II of light microscopy, perhaps due to its low numbers ($<1\%$). The type shown in Fig. 7c, based on its large size, is most probably Type I. Eleocytes of *L. terrestris* are similar to those of nereid polychaetes (Baskin, 1974, see Chapter 3) and of *E. foetida* (Chapron, 1970; Stang-Voss, 1971; Valembois, 1971b). A detailed ultrastructural study of *E. foetida* eleocytes by Lindner (1965) revealed the presence of crystalline bodies resembling ferritin or haemoglobin in the eleocyte inclusions.

7. *Subsidiary or transitional cells (light microscopy)*

Transitional cells, while not a separate cell type, comprise nearly 2% of the total cell population and include all cells showing intermediate characteristics between two cell types. For example, basophils occasionally exhibit varying amounts of acidophilic areas, whereas in others, basophils may show indistinct outlines of basophilic granules of the size found in granulocytes. Both of these are considered to be transitional.

D. *Coelomocytes of other families*

Members of the Megascolecidae and Enchytraeidae often possess, in addition to the basic cell types found in the Lumbricidae (i.e. amoebocytes and eleocytes), two other major cell types, *lamprocytes* and *linocytes* (Table II). In these groups, lamprocytes are often the predominant type, whereas linocytes are usually few in number. It is not clear if these cell types are directly comparable in the two families; however, there are many morphological similarities.

In both families, lamprocytes (Figs 8a–c) are relatively large cells (20–30 μm), flattened or disc-shaped, with round, oval or spindle-shaped outlines. They are non-amoeboid and covered by a thickened external membrane or pellicle. The cytoplasm is filled with clear or slightly yellow, non-lipoid, often refringent vesicles. In *Enchytraeus hortensis* the vesicles are homogeneous (Goodrich, 1897; Hess, 1970). However, in the megascolecids, *Acanthodrilus annectans* and *Octochaetus multiporus*, the vesicles, which are colourless, contain a secondary granule, slightly greenish in colour and highly refringent (Benham, 1901). In *Pheretima indica* and *P. sieboldi* (also megascolecids), the vesicles either contain a similar secondary granule or are stratified into discrete zones (Kindred, 1929; Ohuye, 1937). In *E. hortensis*, Goodrich (1897) found that some lamprocytes, instead of the usual condition of floating freely in the coelomic fluid, are attached to the peritoneum of the body wall and septa by a narrow stalk. In megascolecids this attached condition apparently does not occur. Both Ohuye (1937) and Benham (1901) found many intermediate stages between lamprocytes and eleocytes and proposed a developmental relationship between them. In

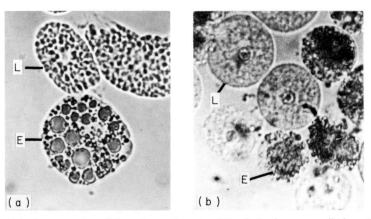

Fig. 8(a). Lamprocytes (L) and an eleocyte (E) of *Enchytraeus albidus*. Living preparation. (Reproduced with permission from Fänge, 1950). × 900.
Fig. 8(b). Lamprocytes (L) and eleocytes (E) of *Bryodrilus* sp. Living preparation. (Reproduced with permission from Fänge, 1974). × 900.

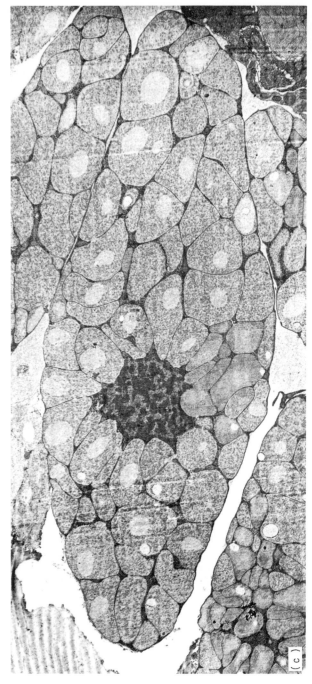

Fig. 8(c). Mucocyte (lamprocyte) of *Enchytraeus fragmentosus*, showing spindle shape and typical granulation. (Reproduced with permission from Hess, 1970). × 5500.

various species of *Enchytraeus* no intermediate stages have been noted; however, Fänge (1974) considers both types to be trephocytic and suggests that no strict distinction should be made between them. In *E. fragmentosus* the lamprocytes have been termed "mucocytes", owing to the presence of acid mucopolysaccharides in the vesicles (Hess, 1970).

Linocytes of the Megascolecidae and Enchytraeidae also exhibit certain differences as well as similarities. The linocytes of *E. hortensis*, originally called "thread corpuscles," contain as the distinguishing characteristic, a colourless refringent body, shaped like a thick disc or truncated cone which appears to be a tightly coiled thread (Goodrich, 1897). On contact of the cell with various solutions, particularly hypotonic media, the inclusion unwinds to reveal a homogeneous thread-like structure. In enchytraeids, the linocytes are sometimes attached to the septa or body wall, as are the lamprocytes and, except for the presence of the thread-like inclusion, are otherwise apparently identical. In megascolecids, however, lamprocytes and linocytes are morphologically dissimilar. Linocytes are partially or completely filled (beside the thread-like inclusion) with one or more non-staining vacuoles, often giving the cell a "honey-comb" appearance. Other vacuolated cells are also present which are similar in appearance but which lack the thread-like inclusion, and are presumed to be precursors to linocytes (Benham, 1901). In all species, the thread-like inclusions react positively with lipid stains and solvents (Goodrich, 1897; Ohuye, 1937) and in *Enchytraeus albidus* are strongly birefringent, with optical properties similar to those of the myelin sheath of vertebrate nerves (Fänge, 1950). Fänge has suggested that the inclusion consists of lamellae of phospholipids, possibly lecithin, and the uncoiled thread-like structures which appear after contact with hypotonic media are myelin figures.

In the Enchytraeidae, coelomocyte morphology has been studied in detail only in *E. albidus* and *E. fragmentosus*. However, Nielsen and Christensen (1959, 1961), in taxonomic studies of 21 genera and over 90 species, have given brief descriptions of the coelomocytes, and whereas morphological information is insufficiently detailed to classify the cells, clearly the coelomocytes of this family are highly varied.

In an earlier study on the megascolecid species *Ocnerodrilus lacuum* and *Dichogaster crawi*, Eisen (1900) reported the occurrence of a coelomocyte type which he called a "mucocyte", and which in the latter species was described as containing foam-like cytoplasm. In *O. lacuum* he considered the mucocyte to be derived from specialized "lymphatic tissue" associated with septal glands in segments VIII–X; such glands are similar to those described in *Pheretima* by Kindred (1929) (see Section IIIA). In *Drawidi hattamimizu* (family Moniligastidae), Ohuye (1934) found, along with amoebocytes and eleocytes, lamprocyte-like cells similar in appearance to those found in

Enchytraeus and *Pheretima* and a vacuolated cell which he called "linocyte". From the above it is obvious that the coelomocytes of the majority of oligochaete families either remain uninvestigated or have been treated very briefly, usually as part of a larger systematic study. Coelomocytes of several species of Tubificidae and Naididae have been briefly mentioned by Stephenson (1930) and of Glossoscolecidae by Eisen (1900); this information is included in Table II.

E. Blood cells

Earlier studies on blood cells have been summarized in a review by Stephenson (1930), its relative brevity reflecting the scarcity of information. Vejdovsky (1884), describing the aquatic worms *Aelosoma*, *Criodrilus*, *Lumbriculus* and *Rhynchelmis*, gave no details of cell morphology, but reported that all their blood cells are permanently attached to the vessel walls by thread-like processes. In *Tubifex*, some cells are similarly attached, but many are freely floating. DeBock (1900) described the freely floating cells of aquatic worms as usually being without pseudopodia, variable in size, and often containing brown or black granules. In enchytraeid worms the attached cells may be either closely adherent to the vessel walls or anchored by fine processes, allowing them to be moved by currents of blood (Nusbaum, 1902). Gungl (1905) found that the cells in members of the Lumbricidae are round or oval with finely granular protoplasm and often resemble coelomocytes. Lison (1928) described blood cells as being small, with many fine granules; some, in a "quiescent" state, had short, thorn-like surface projections, while others, in an "active" state, had large ruffled hyaline pseudopodia.

In more recent ultrastructural studies of *E. foetida*, Burke (1974a) reported only one cell type, containing pale ground cytoplasm and vesicles with dense condensations. Chapron (1970) described blood cells as possessing relatively long pseudopodia interlaced with microfibrils. The cytoplasm contains many free ribosomes, some rough endoplasmic reticulum, and mitochondria with a particularly dense matrix. According to Valembois (1970, 1971b), blood cells may be similar in appearance to coelomocytes of the splanchnopleural cell line, particularly lymphocytoid or macrophage-like cells (Figs 9a,b,c) (see Table I).

V. Functions of coelomocytes

A. Introduction

Oligochaete coelomocytes have varied functions, including the ability to recognize and eliminate foreign materials, primarily by phagocytosis and encapsulation, and to participate in clotting, wound healing, and some aspects of nutrition and excretion (Table VI). Earthworm coelomocytes have been widely studied in relation to the evolution of cell-mediated immunity, in particular to their role in graft rejection and to the related phenomena of antigen recognition, adoptive transfer and immune memory. Other coelomocyte properties related to immune function, such as rosette formation, mitogen responses and chemotaxis have also been investigated, and are included in this section.

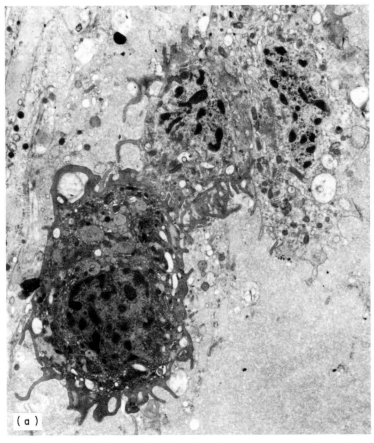

Fig. 9(a). Blood cells of *Eisenia foetida* within the dorsal vessel. (Reproduced with permission from Valembois, 1970). × 5400.

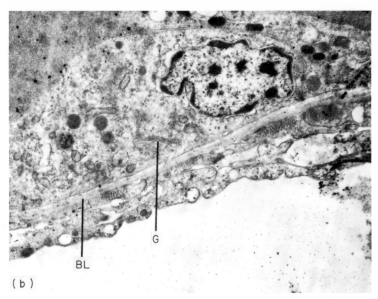

Fig. 9(b). Blood cells of *E. foetida* with macrophage-like appearance, showing Golgi body (G) and basal lamina of blood vessel (BL). (Reproduced with permission from Valembois, 1970). × 11 500.

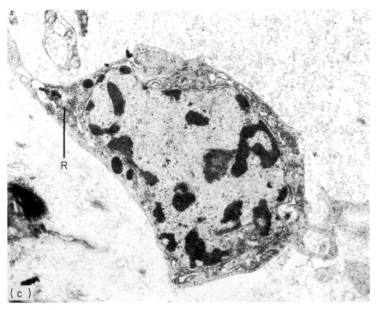

Fig. 9(c). Blood cell of *E. foetida* with appearance of lymphocytoid coelomocyte, containing numerous free ribosomes (R). (Reproduced with permission from Valembois, 1970). × 8500.

B. Nutrition

The function of chloragogen tissue has been compared to that of the vertebrate liver (Avel, 1959; Valembois, 1971b), and the free cells, or eleocytes, regarded as nutritive or trophic (Cuénot, 1891; Liebmann, 1942; Stang-Voss, 1971; Valembois, 1971b). The chloragosomes and other eleocyte inclusions contain large amounts of lipids, protein and glycogen, presumably absorbed from the gut and associated blood vessels (or formed as metabolic products from absorbed substances) with which the chloragogen tissue is closely associated (Cuénot, 1891; Liebmann, 1942; Stein and Cooper, 1978). According to Liebmann (1942), eleocytes migrate into all organs where the cell contents are released and used as nutritive material. Eleocytes are often found in close association with developing eggs (Hess, 1970) and growing embryos (Freudwiler, 1905; Issel, 1906; Liebmann, 1926, 1942; Bell, 1943). Eleocytes also release their granules or disintegrate while in the coelomic fluid, with the various substances subsequently taken up and metabolized by the other coelomocytes (Cameron, 1932; Liebmann, 1942; Valembois, 1971b; Stein et al., 1977). Valembois (1971c), found that even during transplant rejection, products of the destroyed body wall grafts on E. foetida were incorporated into host chloragogen cells and subsequently used as nutritive material.

C. Excretion

Chloragogen tissue and eleocytes have an excretory function, since they may absorb waste material from the blood and release it (contained within the eleocytes) into the coelomic fluid. After eleocyte dissolution or excretion of waste products from them, the material is either removed by the nephridia or taken up by amoebocytes and subsequently discharged through the dorsal pores (Kükenthal, 1885; Cuénot, 1898; Kollmann, 1908). Observations on nitrogen metabolism have partially supported this hypothesis (Semal Van-Gansen, 1956) in that chloragogen tissue has been found to participate in the formation of ammonia (Cohen and Lewis, 1949, 1950; Needham, 1957) and urea (Cohen and Lewis, 1950; Abdel-Fattah, 1955). Chloragogen tissue may also function as a detoxifying tissue for worms living in soils containing high concentrations of heavy metals. Ireland and Richards (1977) have found that worms can tolerate soils heavily contaminated with lead by concentrating the metal in "debris organelles" of chloragogen cells and subsequently excreting it after eleocyte degeneration.

TABLE VI. Coelomocyte functions (see text for: mitogen responses, adoptive transfer and rosette formation).

Function	Coelomocyte type	Species studied	References
Phagocytosis	amoebocytes	species from many families	Metchnikoff (1891), Keng (1895), Cuénot (1898),[a] Kollmann (1908), Kindred (1929),[b] Stephenson (1930), Cameron (1932), Liebmann (1942), Duprat and Bouc-Lassalle (1967), Hess (1970), Stang-Voss (1971), Valembois (1971b), Burke (1974a,b), Stein et al. (1977)
	eleocytes	Eisenia foetida	Duprat and Bouc-Lassalle (1967)
Encapsulation and formation of "brown bodies"	amoebocytes eleocytes[g]	many species (various families)	Metchnikoff (1891), Keng (1895), Cuénot (1898), Kindred (1929), Cameron (1932), Duprat and Bouc-Lassalle (1967), Poinar and Hess (1970)
Wound closure (formation of wound plug)	all coelomocytes	Lumbricidae	Cameron (1932), Duprat and Bouc-Lassalle (1967), Chapron (1970), Stang-Voss (1971),[c] Burke (1974a,b).
Wound healing and regeneration (general)	amoebocytes	Eisenia foetida	Duprat and Bouc-Lassalle (1967), Valembois (1971b)
	eleocytes	Eisenia foetida Lumbricus terrestris	Liebmann (1942, 1943), Duprat and Bouc-Lassalle (1967)
Formation of substratum for regenerating epidermis during wound healing	hyaline amoebocytes	Eisenia foetida	Burke (1974a,b)
Graft rejection	amoebocytes "splanchnopleural macrophage"[d]	Lumbricus terrestris Eisenia foetida	Hostetter and Cooper (1972) Valembois (1971b,c)
	Type I granulocyte[d] hyaline amoebocytes "small basophil"[d]	Lumbricus terrestris Eisenia foetida Eisenia foetida	Linthicum et al. (1977a) Chateaureynaud-Duprat (1971) Roch et al. (1975), Valembois and Roch (1977)

Function	Cell type	Species	References
Cytotoxicity	unspecified	*Lumbricus terrestris*	Cooper (1973a)
	eleocytes[e]	*Eisenia foetida*	Chateaureynaud-Duprat and Izoard (1977)
	amoebocytes[f]	*Eisenia foetida*	
Possible blood cell precursors	amoebocytes	Lumbricidae (several species)	Cuénot (1898), Cameron (1932), Valembois (1971a)
		Pheretima indica	Kindred (1929)
Possible precursor cells for epidermal basal cells and inter-muscular granule-containing cells	hyaline amoebocytes	*Eisenia foetida*	Burke (1974a,b)
Haemoglobin production	haemoglobin-producing cell	*Eisenia foetida*	Valembois (1970, 1971b)
Haemopoiesis	eleocytes	*Eisenia foetida*	Stang-Voss (1971)
Nutritive ("trephocytic")	eleocytes	species from many different families	Cuénot (1891), Liebmann (1927, 1931, 1942), Cameron (1932), Stang-Voss (1971), Valembois (1971b), Burke (1974a,b), Stein and Cooper (1978)
Association with developing eggs (possible nutritive function)	lamprocytes	*Enchytraeus fragmentosus*	Hess (1970)
Excretory	eleocytes	species from many families	Cuénot (1898), Stephenson (1930), Abdel-Fattah (1955), Semal-Van Gansen (1956), Roots (1960)

[a] Young amoebocytes only; [b] hyaline amoebocytes only; [c] primarily "transport-form" amoebocytes only; [d] subtype of hyaline amoebocytes; [e] non-specific, naturally occurring; [f] specific, acquired; [g] probably passively involved.

D. *Reactions to injury: clotting, wound-healing and regeneration*

Injuries to the body wall of earthworms elicit coelomocyte responses that vary with the type of injury. Wounds which penetrate the coelom are usually sealed off by a plug of aggregated coelomocytes. In most species, with only a few exceptions (see below), the fluid phase of coelomic fluid and of blood does not coagulate; rather, the coelomocytes aggregate together to form a "wound plug", a mechanical barrier which prevents further loss of fluid and cells (Cuénot, 1891; Goodrich, 1897; Kindred, 1929; Grégoire and Tagnon, 1962; Burke, 1974a). The wound plug consists of all types of coelomocytes, although amoebocytes predominate; eleocytes, when present, are often fragmented and their contents phagocytized by other cells of the plug (Liebmann, 1942; Chapron, 1970; Burke, 1974a). Most amoebocytes comprising the wound plug contain many electron-dense, lysosome-like granules and peripheral microfilaments, and have lobopodia (Burke, 1974a), and are thus structurally like neutrophils (i.e. Type I granulocytes) (Linthicum *et al.*, 1977a). A few hours after plug formation, these cells become flattened and spread over the wound surface to become a provisional wound covering (Fig. 10), forming the stratum on which the regenerating epidermis later spreads as healing progresses (Chapron, 1970; Valembois, 1971c; Burke, 1974a,b). Similarly, damaged or severed blood vessels are mechanically closed by the formation of a plug of blood cells (Burke, 1974a).

In at least three species of the Megascolecidae, *Acanthodrilus annectans*, *Octochaetus multiporus* (Benham, 1901) and *Pheretima sieboldi* (Ohuye, 1937), the fluid phase of the coelomic fluid clots. In *P. sieboldi*, Ohuye (1937) has observed that clotting can occur in the absence of coelomocytes. Cameron (1932) stated that the coelomic fluid of lumbricid worms coagulates; however, we have never observed this in *L. terrestris*, nor does it appear to occur in *E. foetida* (Burke, 1974a).

Shallow wounds in the body wall (i. e. not penetrating into the coelom) are also filled, within a few hours, by a wound plug of coelomocytes which are similar in ultrastructure to those in penetrating wound plugs (Burke, 1974a). Cameron (1932) observed that 1 h after wounding, coelomocytes (primarily hyaline amoebocytes), increased in number within the coelom directly under the wound site and 2 h later, small areas of the coelomic epithelium began to proliferate. Six hours after wounding, coelomocytes migrated into the body wall near the site of injury and 24 h later, a plug of coelomocytes, now containing some acidophils (granular amoebocytes) extended from the coelom into the wounded area. Later, at about two days, the epidermis began to grow over the wound when the plug still consisted chiefly of hyaline amoebocytes and granular amoebocytes, the former having phagocytized much of the dead tissue. Spindle-shaped cells, believed to be undifferentiated cells of non-coelomic origin, also appeared within the plug. By ten days, healing was almost complete and the body-wall tissue nearly normal in structure.

Liebmann (1942, 1943) has emphasized the importance of eleocytes in wound-healing and regeneration and has proposed that some component of eleocytes, possibly guanine (reported as present by Willem and Minne, 1900), is essential for regeneration to occur. However, Moment (1974) has reported that the presence or absence of eleocytes has no effect on the course of regeneration, and studies by Abdel-Fattah (1955), Semal Van-Gansen (1956) and Roots (1960) have indicated that guanine is not present in chloragogen tissue.

E. Immune recognition

The ability to distinguish between self and non-self is a fundamental

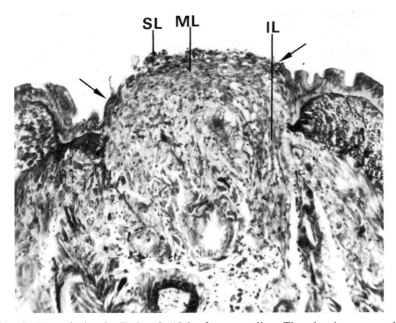

Fig. 10. Wound plug in *E. foetida* 12 h after wounding. The plug is composed of three layers of coelomocytes: (1) a surface layer (SL) of cells that are continually sloughed; (2) a middle layer (ML) of densely-packed cells lying parallel to the wound surface; and (3) an inner layer (IL) of coelomocytes elongated toward the wound. Arrows indicate the edge of the migrating epidermis. (Reproduced with permission from Burke, 1974b). × 270.

property of most, if not all, animals (Burnet, 1974; Cooper, 1977). A specialized form of this property, *immune recognition*, occurs as an essential first step in all immune responses studied in vertebrate leucocytes and to a lesser extent in those of invertebrates, emphasizing specificity in its relationship to cell-surface properties. Immune recognition is apparent in the many defence functions of earthworm coelomocytes, although rarely studied as a separate event. Both phagocytosis and encapsulation require "not-self" recognition and, while phagocytosis is generally regarded as non-specific since antibody (which is not present in invertebrates) is not required, earthworm coelomocytes do display a certain specificity in the uptake of bacteria and of allogeneic and xenogeneic sperm (see section on phagocytosis) (Cameron, 1932).

Transplantation studies have also revealed a degree of selective coelomocyte responsiveness, in that invasion and destruction of foreign tissue grafts (a coelomocyte-mediated process) occurs more quickly and vigorously with inter-specific (xenogeneic) than with intra-specific (allogeneic) grafts, and self-tissue grafts are not destroyed (Duprat, 1964, 1967; Cooper, 1968, 1969a,b; Valembois, 1970).

Coelomocytes are also capable of cytotoxicity (Cooper, 1973a) and of binding sheep erythrocytes (SRBC) by forming rosettes (Cooper, 1973b; Toupin and Lamoureux, 1976a; Cooper and Perussia, in preparation). Binding appears to be specific, since rosettes do not form with mouse, human or chicken erythrocytes, and only in low percentages with ox erythrocytes. Rosettes form primarily with 25–45% of small, non-adherent coelomocytes, having basophilic, non-granular cytoplasm and a high nucleo-cytoplasmic ratio. Specificity of binding suggests that coelomocytes possess recognition units or receptors that match or fit antigenic determinants on the SRBC.

F. Phagocytosis

Coelomocytes play an essential role in defence against invading microorganisms. The coelomic cavity connects directly to the external environment via the dorsal pores and nephridiopores, and consequently it often contains a number of bacteria, fungal spores, protozoa and nematodes (Keng, 1895; Cuénot, 1898; Cameron, 1932; Marks and Cooper, 1977). Phagocytosis in the oligochaetes was first investigated in 1891 by Metchnikoff, who described the earthworm coelomocyte as "among the most active of phagocytes . . . (which) devours with the greatest eagerness all foreign bodies which come in its way". Cuénot (1898), studying the relative phagocytic activity of different coelomocyte types, found that small hyaline amoebocytes were the most active, granular amoebocytes less so, and eleocytes inactive.

Cameron (1932) investigated phagocytosis in lumbricid worms *in vivo* and *in vitro* using a variety of foreign particles. India ink, carmine, colloidal iron, and dust and coal-particle suspensions were all taken up most avidly by all sizes of basophils (hyaline amoebocytes), less so by acidophils (granular amoebocytes), and not at all by eleocytes. Coelomocytes ingested droplets of olive oil and milk fat only slightly. Basophils, but no other coelomocyte types, were moderately phagocytic for the sperm of rabbits, mice, and other worm species, but were inactive against homologous sperm. Many bacterial species were vigorously phagocytized, again most strongly by basophils, less so by acidophils, but not at all by eleocytes; however, some species, such as *Bacillus tetanus*, *B. welchii*, *Corynebacterium diphtheriae*, and the haemolytic strain of *Streptococcus pyogenes* were more slowly ingested. Some differences among earthworm species were also noted; coelomocytes of *Lumbricus terrestris* were more active than those of *Allolobophora longa* or *Octolasium cyanum* for some types of bacteria.

More recently, Stein *et al.* (1977), measuring *in vivo* phagocytosis of carbon particles injected intracoelomically into *L. terrestris*, found that approximately 50% of all coelomocyte types took up carbon, except eleocytes, which were inactive. Carbon amounts varied among cell types, with basophils and neutrophils containing greater amounts than either acidophils or granulocytes (Fig. 11). *In vitro* uptake of yeast was also measured, varying the concentration of coelomic fluid in the incubating medium to determine effects of factors acting as opsonins on phagocytosis. Phagocytosis by basophils, acidophils and granulocytes appeared to be unaffected by the presence or absence of coelomic fluid; basophils and neutrophils were again more phagocytic than acidophils or granulocytes. Only neutrophils were more phagocytic when coelomic fluid was present, indicating a possible (though small) opsonizing effect. Type I chloragogen cells (eleocytes) were never found to contain yeast, although Type II chloragogen cells occasionally enclosed a few (Stein and Cooper, in preparation).

G. Encapsulation

Foreign objects such as protozoa or nematodes, too large to be phagocytized, are inactivated by encapsulation. Metchnikoff (1891) was again one of the earliest investigators to study this process in oligochaetes, describing the encapsulation of larvae of *Gordius* (horsehair worms) by peritoneal cells of *Nais proboscideae* (which lacks free coelomocytes) and of cotton threads, gregarines and nematodes by coelomocytes of *L. terrestris*. The capsules were composed of two layers; an inner, non-cellular, possibly chitinous layer, which Metchnikoff (1891) believed to be secreted by the parasite, and an outer cellular layer consisting of immobile, highly flattened and occasionally deeply-pigmented coelomocytes. The capsules were sometimes thickened and formed of many layers of cells, or at other times, remained thin and with a structure similar to that of connective tissue. Cuénot (1898), studying the encapsulation of parasites, including the spores and sporozoites of gregarines, considered the inner non-cellular layer to be a coelomocyte secretion.

Cameron (1932), after inserting cotton thread into the coelomic cavity, noted successive stages in capsule formation. Initially, a thin layer of coelomocytes surrounded the thread, followed by the addition of new cell

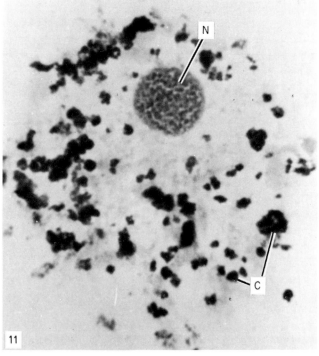

Fig. 11. Neutrophil of *L. terrestris* containing phagocytized carbon particles (C). Cytoplasm is very light and contains a few scattered granules. Nucleus (N) is slightly larger and lighter-staining than other coelomocyte types. Wright's stain. (Reproduced with permission from Stein *et al.*, 1977). × 2000.

layers and this was accompanied by a flattening of the inner layers. After seven days, thin fibrous strands were present throughout the capsule.

The ultrastructure of "brown bodies", structures found primarily in the terminal segments of earthworms and containing encapsulated nematodes, protozoan cysts and debris, was described by Poinar and Hess (1977) (Figs 12a,b). They found the inner non-cellular layer of the capsule to be homogeneous, and proposed that it is a deposit of a non-cellular component of the coelomic fluid, perhaps an agglutinin. The outer cellular layer was composed of at least four coelomocyte types: cells which lack phagosomes and inclusion granules (the innermost cellular layer); cells with many electron-dense bodies, similar in appearance to Type I granulocytes (neu-trophils) of Linthicum *et al.* (1977a); coelomocytes containing large spherical vesicles (possibly inclusion-containing cells, i.e. acidophils); and coelo-mocytes containing large pleomorphic granules with electron-dense zones. Light-microscopic studies have shown that brown bodies frequently contain eleocytes and the central part is often rich in chloragosomes (Duprat and Bouc-Lassalle, 1967; Semal Van-Gansen, 1956).

H. Transplant rejection

1. Introduction

Earthworms have figured prominently in studies concerning the phylo-genetic development of cell-mediated immunity (Valembois, 1963; Duprat, 1964, 1967; Cooper, 1965a,b, 1968, 1969a,b; Cooper and Rubilotta, 1969). Transplantation experiments have demonstrated that earthworms are cap-able of recognizing and rejecting foreign tissue grafts while accepting grafts of self tissue. All grafts, both foreign and self, initially heal, become vascularized and firmly attached to the host. Later, however, foreign grafts show varying degrees of rejection while self-tissue grafts (autografts) heal completely. Of the foreign tissue grafts, xenografts are rejected the most vigorously and usually completely, while allografts are rejected more slowly and often incompletely (Duprat, 1964, 1970; Cooper, 1968, 1969a,b).

2. Role of coelomocytes

(*a*). *Histopathology of graft rejection.* Light-microscopic histological studies conducted during the process of graft rejection have shown that within 24 h of grafting, coelomocytes congregate around and particularly underneath the graft sites and infiltrate the graft matrix. Apparently this occurs as a non-specific response to injury (see section on reactions to injury: clotting, wound-healing and regeneration), since in the case of autografts, such infiltration stops as soon as damaged tissue has been removed and healing is well underway. In the case of xenografts, however, coelomocytes continue to invade the graft tissue and remain until it is completely destroyed. In addition, greater numbers of coelomocytes accumulate under, as well as within, xenografts, as compared to autografts (Cooper, 1969a,b; Hostetter and Cooper, 1972). Host coelomocytes also surround the graft in varying numbers and in the most extreme condition, may completely enclose the entire graft, walling it off from the host in an encapsulation-like process (Hostetter and Cooper, 1972; Parry, 1978). The reason for the variability of this latter response is unclear, but may be related to the extent of injury to the graft-bed of the host by the surgical procedures (Hostetter and Cooper, 1972).

Electron microscopic studies of xenograft rejection (*E. foetida* grafts on *L. terrestric* hosts) have shown the following sequence of events (Linthicum *et al.*, 1977b): (i) Within 1–3 days post-transplantation, the graft matrix is infiltrated by coelomocytes that have been identified as Type I granulocytes (neutrophils, by light microscopy). (ii) By three days post-transplantation, destruction of the inner muscle layer has begun, with changes in muscle fibre organization and the appearance of large vacuolated areas (Fig. 13). Many

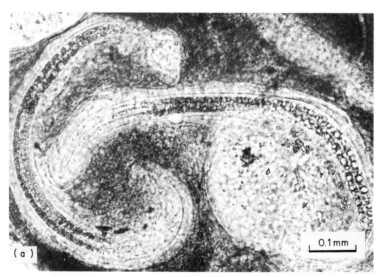

Fig. 12(a). Two specimens of the nematode, *Rhabditis pellio*, coiled within a "brown body" of the earthworm, *Aporrectodea trapezoides*. Note lighter uniform deposit surrounding the nematodes. (Reproduced with permission from Poinar and Hess, 1977).

Type I granulocytes are present, within which are visible large phagocytic vacuoles containing pieces of apparently viable muscle fibre. (iii) By day 5, the Type I granulocytes contain numerous phagosomes, residual bodies and glycogen granules (Fig. 14). (iv) By 11–13 days, Type I granulocytes have migrated into the outer layer of muscle, and almost all muscle tissue has been destroyed (Figs 15, 16). The outer epithelium is still intact. At this stage small lymphocytic (basophilic) coelomocytes are occasionally visible within the graft; however, they appear to be minor scavengers, phago-cytosing small cellular debris but containing no viable muscle fragments.

During graft destruction muscle damage appears to be progressive, i.e. occurs earliest in those muscle layers closest to the coelomic cavity, and is concurrent with the appearance of infiltrating coelomocytes; later, as coelomocytes migrate further into the outer muscle layers, destruction is then evident in these areas. As in the light microscopic studies, ultra-

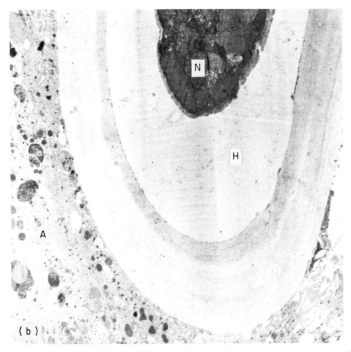

Fig. 12(b). Section through a "brown body" of *A. trapezoides* showing the nematode *R. pellio* (N) surrounded by a homogeneous layer (H) and an outer area consisting of amoebocytes (A). (Reproduced with permission from Poinar and Hess, 1977). × 2200.

structural studies (Linthicum *et al.*, 1977b) have shown that coelomocytes infiltrate not only foreign tissue grafts, but also autografts and the wounded area of sham-operated worms; however, in these instances the reaction is a non-specific, general inflammatory response, related to wound-healing and removal of damaged tissue, with no visible phagocytosis of viable muscle.

Valembois (1968, 1974) has reported a somewhat different situation in xenograft experiments employing *E. foetida* as host. Ultrastructural studies by this author have indicated that in this species destruction of first-set grafts apparently occurs as a result of autolysis, involving synthesis of lysosomal enzymes by the transplanted tissues. Destruction of second-set grafts, however, results from coelomocyte invasion, as in first-set xenografts of *L. terrestris*. Valembois has identified the coelomocytes which infiltrate

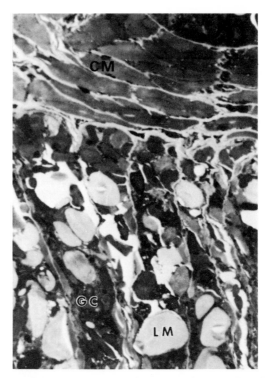

Fig. 13. Early stage (3 days) in rejection of *E. foetida* body wall graft on *L. terrestris* recipient. Coelomocytes (GC) have invaded inner longitudinal muscle layer (LM) of graft, which shows extensive damage, but not outer circular muscle layer (CM), which remains intact. 1 μm section. (Reproduced with permission from Linthicum *et al.*, 1977b). × 440.

the graft as macrophages (= Type I granulocytes) that have their origin in stem cells of the splanchnopleural epithelium.

(b). *Differential coelomocyte responses to grafting and injury.* As mentioned in the previous section (2a) grafting or injury to the body wall of earthworms is accompanied by an increase in the number of free coelomocytes in the coelomic cavity underneath the grafted or wounded area. Hostetter and Cooper (1973) have quantified this response to graft rejection and wound healing by measuring the numbers of coelomocytes directly under autografts, xenografts and wounds at different times during the rejection (or, in the case of autografts or wounds, healing) process, using, as xenografts, *E. foetida* grafts on *L. terrestris* hosts. This study has shown that response to wounds is the most rapid, with coelomocyte numbers rising within 24 h; however, the decline is also rapid, with coelomocyte numbers returning to normal by 72 h. Autografts elicit a weaker response, with coelomocyte counts rising only slightly above those of ungrafted worms and returning to normal by day 3. Response to xenografts is slower, reaching a peak in 3–4 days, and declining only slowly, requiring 7 days to return to normal.

At 20°C, the temperature at which the study was conducted, xenograft rejection was completed in a mean time of 17 days, approximately 10 days beyond the decline in coelomocyte numbers under the graft. If a second

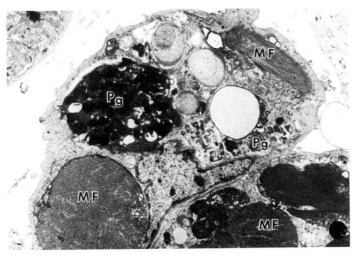

Fig. 14. Type I granulocyte (neutrophil) within xenograft, 5 days post-transplantation, containing large fragments of muscle fibres (MF) and phagosomes (Pg). (Reproduced with permission from Linthicum *et al.*, 1977b). ×9700.

graft was added at this time (17 days post-first xenograft) the second xenograft was rejected in an accelerated mean time of 6 days, indicating a memory component to the rejection response (Winger and Cooper, unpublished data; Hostetter and Cooper, 1973). Counts of coelomocytes taken from under such second grafts or second wounds varied considerably from those associated with first grafts or wounds. Following second xenografts, coelomocyte numbers were 20–30 % higher than after first grafts, peaking 1–2 days after grafting. Second transplants (including autografts) and wounds were also given 5 days after the initial procedure, and results were similar to those found in the 17-day group (see above). Such accelerated and heightened responses to second grafts suggest presence of a memory component in coelomocyte response.

(c). *Adoptive transfer.* Additional experimental evidence for the involvement of coelomocytes in graft rejection comes from the demonstration of adoptive transfer (Duprat, 1967; Bailey *et al.*, 1971; Valembois, 1971a). In these experiments, host *L. terrestris* (A) were first xenografted with *E. foetida*, then coelomocytes were harvested at 5 days post-transplantation and injected into ungrafted *L. terrestris* (B). This second host was then

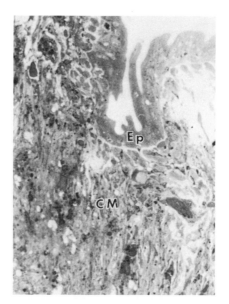

Fig. 15. Thirteen days post-transplantation, coelomocytes have migrated into outer muscle layer (CM) and nearly all muscle has been destroyed. The epithelium (Ep) is still intact. 1 μm section. (Reproduced with permission from Linthicum *et al.*, 1977b). × 450.

xenografted with the same *E. foetida* donor used to induce immunity in A. Because *L terrestris* shows only negligible allograft responses late after transplantation (Cooper, 1969a), no early coelomocyte allo-incompatibility was expected prior to the action of primed coelomocytes against the *E. foetida* graft on *L. terrestris* (B). The second host (B) showed accelerated rejection of its first transplant, thus demonstrating short-term memory and confirming the adoptive transfer of graft rejection capacity by primed coelomocytes. The response is cell-mediated, since transfer of coelomic fluid alone, free of coelomocytes, is ineffective. Coelomocytes from unprimed *L. terrestris* or from *L. terrestris* primed with saline are also unable to transfer the response. How specific this adoptive transfer response is remains to be determined. For example, if we graft a *L. terrestris* host with *E. foetida*, then transfer the coelomocytes and graft the new host with a third species, what is the response, how long does it last, and what is the role of temperature in regulation? Is memory longer lived at lower temperatures?

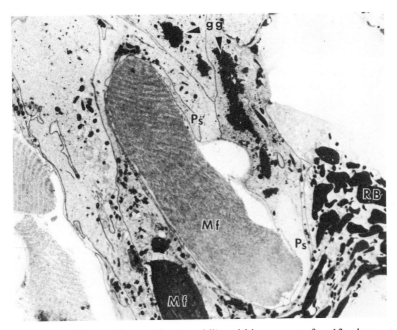

Fig. 16. Type I granulocyte (neutrophil) within xenograft, 13 days post-transplantation. Large pseudopodia (Ps) surround a piece of muscle fibre (MF). Coelomocytes contain glycogen granules (GG) and residual bodies (RB). (Reproduced with permission from Linthicum *et al.*, 1977b). × 14 250.

I. Behaviour in culture

1. Separation by velocity sedimentation

Earthworm coelomocytes are heterogeneous morphologically and participate in diverse immunological functions. Which specific cells mediate these responses is controversial, thus we have developed a procedure for separating coelomocytes by velocity sedimentation at unit gravity (Cooper et al., 1980). This separation technique, widely applicable in mammalian systems, selects cells primarily on the basis of size and enriches for at least two distinct coelomocytic types, basophils and neutrophils, which participate in graft rejection (Linthicum et al., 1977b). After purifying distinct coelomocytes, we now have a method for subjecting them to various immunobiological tests.

When coelomocytes are applied as a suspension to a step gradient of 7–30 % calf serum in buffered medium, allowed to sediment for 1·5–2 h at 4°C, then recovered in specific fractions, they are broadly distributed into two major peaks. Basophils comprise the first peak, whereas the second consists almost exclusively of granulocytes and neutrophils. Significant enrichment (two- to three-fold) has been observed and highly purified populations obtained in certain fractions. For example, basophils have been enriched almost two-fold in slowly sedimenting fractions and both granulocytes and neutrophils three-fold in rapidly sedimenting fractions. In absolute terms, the purity of basophils and of granulocytes is greater than 80 % in the most enriched fractions. These cells show the typical structure as observed in previous light microscopic preparations. Acidophils constitute 6·6 % of the unseparated population and are mainly found in rapidly sedimenting fractions. Chloragogen and transitional cells in the unseparated fractions are present in low frequency as has been found previously, however, upon separation, we have observed no significant enrichment. We confirmed the ultrastructural appearance of cell types using the following terminology: Types I and II lymphocytic coelomocytes correspond to basophils in light microscopy, ranging in size from about 4–9 μm, with numerous membrane limited vacuoles, smooth and rough endoplasmic reticulum, Golgi vesicles and pseudopodial formations. Granulocytes and neutrophils were variable in size and granulocytes ranged from 9–15 μm. The cytoplasm contained dark granules, endoplasmic reticulum, and Golgi.

2. Effects of stimulation by mitogens and transplantation antigens

The origin of the coelomocytes which accumulate under body-wall grafts or injuries, and which ultimately participate in graft rejection and/or wound-healing, has never been decisively determined. Heightened proliferation of the epithelial lining of the coelomic cavity and septa occurs during wound-healing (Cameron, 1932), and possibly during grafting (Hostetter and Cooper, 1974), and may be the principle source of these cells. The division of free coelomocytes, either normally or as the result of stimulation by wounding or grafting, has been suggested as an additional source of cells, but the mitotic capability of free coelomocytes has yet to be definitely proven. Cuénot (1898) described the division of free coelomocytes by a process that he considered to be "amitosis"; Kollmann (1908) and Cameron (1932) later reported observations of cell division. Other investigators, however, have observed no mitotic figures among free coelomocytes (Kindred, 1929; Liebmann, 1942; Hostetter and Cooper, 1974; Stein et al., 1977).

Mitogens such as bacterial lipopolysaccharides (LPS) and the plant lectins concanavalin A (Con A) and phytohaemagglutinin (PHA) have been widely used to study the proliferative responses of vertebrate lymphocytes (Ling, 1968; Möller, 1972; Oppenheim and Rosenstruck, 1976). Similarly, mitogens have been employed to investigate the proliferative responses of earthworm coelomocytes, and these responses compared to those resulting from stimulation by transplantation antigens (Lemmi, 1975a,b; Roch et al., 1975; Toupin and Lamoureux, 1976b; Toupin et al., 1977; Roch, 1977; Valembois and Roch, 1977; Roch and Valembois, 1978).

(a). Con A. Initial investigations by Roch et al. (1975) using Con A, demonstrated that the coelomocytes of E. foetida are capable of increased DNA synthesis in response to mitogen stimulation. The number of responding cells is low, however, and autoradiographic studies have shown that they constitute less than 1 % of the total coelomocyte population. Peak incorporation of ^{3}H-thymidine was found, in the initial study (Roch et al., 1975), to occur after 5 days incubation with the mitogen, but a subsequent study (Roch, 1977) indicated the highest incorporation at 4 days, with a stimulation index (radioactivity of mitogen-stimulated cells relative to non-stimulated cells) of 4·7.

Heterogeneity in coelomocyte response was found by separating coelomocytes into adherent and non-adherent fractions, then labelling each fraction with fluoresceinated Con A (FITC Con A) (Roch and Valembois, 1978). All coelomocytes bound Con A on their surfaces and initially both populations exhibited the same type of binding pattern, a uniform distribution over the entire cell surface. Non-adherent coelomocytes, however, bound greater concentrations of mitogen than adherent, and by measuring uptake of ^{3}H Con A, a maximal number of 1×10^5 molecules were estimated to be bound to the surfaces of small non-adherent cells. With time, the uniform distribution pattern changed, however, the particular configuration varied with the cell type. Thirty minutes after labelling, adherent cells displayed small patches of fluorescence, distributed randomly over the cell surface. Two hours after labelling, surface patches disappeared and the cytoplasm was uniformly fluorescent. In the non-adherent population, larger cells (constituting 80% of this group) showed patching similar to that of adherent cells. Smaller non-adherent cells (20% of this group) also showed initial patching, but later, at 1 h, 30% of the small cells (i.e. 6% of the non-adherent cells) formed a fluorescent cap, varying in size from approximately half the cell surface to small caps of less than one-fourth the surface. After 90 min, the cap was internalized and the cytoplasm became uniformly fluorescent. Smaller cells (4–6 µm) tended to develop smaller caps, while larger cells (7–9 µm) usually had half-surface caps. The cells which produced smaller, more concentrated caps were characterized by sparse cytoplasm containing only free ribosomes. During the process of surface redistribution and internalization of mitogen, a Golgi apparatus and vesicular components appeared. Since internalized mitogen appeared to be associated with lysosome-like vesicles, the development of these various cell organelles might be related to the endocytosis of Con A binding sites.

Formation of the larger, hemispherical caps was not considered to be a "true" capping, but another type of receptor redistribution. Cells with such caps had more abundant cytoplasm than smaller cells, and contained a well-developed Golgi apparatus and endoplasmic reticulum. Redistribution of binding sites to form the cap induced formation, within the cap, of long (3–7 µm) villi containing cytoplasmic organelles. Using scanning electron microscopy, the Con A treated, non-adherent coelomocytes were found to consist of four types, differing in size and in type of surface projections (Roch and Valembois, 1978).

Mitogen inhibition studies, employing a number of different sugars, showed that methyl-α-D-mannose, methyl-α-D-glucose and D-fructose competitively inhibited the binding of Con A to coelomocytes. Methyl-α-D-mannose was the most active, and prevented Con A binding at lower concentrations than either methyl-α-D-glucose or D-fructose (Roch et al., 1975; Roch and Valembois, 1978). The presence of protein in the incubating medium also reduced mitogen binding but was apparently a non-specific effect, since a number of unrelated proteins were equally effective (Roch and Valembois, 1978).

(b). PHA. Toupin and Lamoureux (1976b) and Toupin et al. (1977) demonstrated increased DNA synthesis by L. terrestris coelomocytes in response to PHA stimulation. Using coelomocyte populations separated into adherent and non-adherent fractions, only the non-adherent fraction incorporated ³H-thymidine. For this to occur, however, a portion of the adherent population, designated "trypsin resistant", was necessary.

Maximum thymidine incorporation occurred after four days of incubation, with a stimulation index of approximately five. Roch (1977) found a different type of PHA response in *E. foetida*, in which large adherent cells showed a higher stimulation index (7·8) than small non-adherent cells (1·7), the former reaching peak incorporation at five days and the latter at three days.

(*c*). *LPS.* Information on coelomocyte response to LPS is sparse, consisting of a brief study by Roch (1977) which nevertheless demonstrated that small numbers of coelomocytes incorporate ^{3}H-thymidine in the presence of LPS, with a stimulation index of 3·6. Only small non-adherent cells responded; large adherent coelomocytes and chloragogen cells (eleocytes) were inactive.

(*d*). *Transplantation antigens.* Lemmi (1975a,b) demonstrated the *in vivo* synthesis of DNA, by ^{3}H-thymidine incorporation, in coelomocytes of xenografted *L. terrestris* at various times post-grafting, with a peak incorporation at four days. Roch *et al.* (1975) investigated *in vitro* DNA synthesis by *E. foetida* coelomocytes following wounding, allografting and xenografting. Peak incorporation of ^{3}H-thymidine occurred 4 days after all 3 operations, but was greatest after allografting (with an estimated stimulation index, S.I., of approximately 15), less after xenografting (S.I. $\cong$ 5) and least after wounding (S.I. $\cong$ 4). Responses following second grafts or wounds were more complex with a peak at day 2 for xenografts (S.I. $\cong$ 4), at days 2 and 6 for allografts (S.I. $\cong$ 4), and at day 6 for wounds (S.I. $\cong$ 3). Using autoradiographic techniques, xenografting (particularly second sets) was found to simulate DNA synthesis in greater numbers of small coelomocytes ($< 10\ \mu m$) than in larger cells ($> 10\ \mu m$).

Subsequent extension of the autoradiographic work (Valembois and Roch, 1977) indicated that, at least in terms of DNA synthesis, stimulation of coelomocytes by both wounding and grafting is greater than that which occurs with mitogens. Four days after xenografting, 13% of all coelomocytes incorporated ^{3}H-thymidine *in vitro*, and after second xenografts less (6%) are labelled. After first wounds 6·5% are labelled, and after second wounds 4%. Small cells appear to be stimulated more by grafting than by wounding. In normal worms (no grafts or wounds) larger coelomocytes, characterized by abundant cytoplasm, a few free ribosomes, moderate amounts of endoplasmic reticulum and vesicular elements, and large pseudopods, incorporated ^{3}H-thymidine more frequently than did small coelomocytes characterized by a high nucleo-cytoplasmic ratio and many free ribosomes. After second xenografts, however, the situation is reversed, with more smaller cells labelled than large. Coelomocytes do respond to mitogens, wounds and grafts, responses that show some degree of specificity, confirming the earthworm's primitive immunologic capacity.

3. Chemotaxis

Coelomocytes of *L. terrestris* were found by Marks *et al.* (1979) to respond *in vitro* by directional migration to both bacterial and foreign tissue antigens. In the case of bacteria (*Aeromonas hydrophila* and *Staphylococcus epidermidis*), the magnitude of the chemotactic response was in direct proportion to the bacterial concentration. Responses to body wall tissues from *E. foetida*, *Pheretima* sp. and *Tenebrio molitor* (an insect) were found to be highest toward *E. foetida*, moderate toward *Pheretima* sp., and least toward *T. molitor*, an inverse ratio to phylogenetic relatedness. The *Eisenia* chemotactant was postulated to be a small molecular-weight protein, since it was dialyzable and its activity destroyed by heat. A migration-inhibition factor appeared to be present in *L. terrestris* body-wall tissue, since placing both *E. foetida* and *L. terrestris* tissue together in the chemotactant chamber resulted in reduced migration from that found with *E. foetida* tissue alone. Of the responding coelomocytes 92–94% were found to be neutrophils (Type I granulocytes).

VI. Summary and concluding remarks

Oligochaetes possess both a blood vascular system and coelom, each with separate cell populations. Although coelomocytes have been studied in considerable detail in some species, relatively little is known about blood cells, nor is it known if there is any relationship between the two cell populations. Coelomocytes are believed to be derived from the epithelial lining of the coelomic cavity, or in some species (primarily Megascolecidae), from specialized structures, the "lymph glands". Coelomocyte types vary among the different oligochaete families. Those of Lumbricidae may be sub-divided into two major categories, amoebocytes and eleocytes. These two coelomocyte groups differ not only in their sites of origin, but also in morphology and function. The coelomocytes from families other than the Lumbricidae have not been studied extensively, although a few species from the Megascolecidae, Moniligastridae and Enchytraeidae have received some attention. In these species, two other coelomocyte types in addition to amoebocytes and eleocytes have been identified, namely lamprocytes and linocytes. In the Megascolecidae, these two latter types appear to be related developmentally, and while nothing definite is known about their function, perhaps, like eleocytes, they may be trephocytic.

The coelomocytes of *Lumbricus terrestris* have been studied in detail in our laboratory at the light microscope level (Stein *et al.*, 1977), by cytochemistry (Stein and Cooper, 1978), and with transmission electron microscopy (Linthicum *et al.*, 1977a). The coelomocytes of *L. terrestris* appear to be similar, with minor exceptions, to those of other Lumbricidae. By light microscopy, five major cell types have been identified, consisting of basophils, acidophils, neutrophils and granulocytes (all amoebocytes), and chloragogen cells (eleocytes). Both the acidophil and chloragogen cell groups contain two subgroups. Acidophils are usually granular cells, and occasionally discharge their granules. All cell types, with the exception of chloragogen cells, often produce elaborate and pleomorphic pseudopodia.

Cytochemical techniques have revealed the biochemical composition of the major cell types. The enzymes acid phosphatase and β-glucuronidase are present in all cells but are especially abundant in basophils and neutrophils; presence of these enzymes correlates well with phagocytic activity. Peroxidase is present only in eleocytes and then in low concentrations. The cytoplasmic basophilia of basophils is due primarily to ribonucleic acid, and they also contain large deposits of glycogen. Neutrophils and chloragogen cells also enclose glycogen but in lesser amounts. The predominant granules of acidophils and granulocytes are composed of a basic protein and a neutral mucopolysaccharide or glycoprotein. A second granule population, present in low numbers in acidophils and granulocytes, but in larger numbers in basophils and neutrophils, is small and lipid-positive and may, in part, represent lysosomes. Lipid is especially abundant in the vesicles and granules of chloragogen cells. Some granules of chloragogen cells also contain ferrous and ferric iron and a substance with pseudoperoxidase activity. Acid mucopolysaccharides are present in chloragogen cells and in no other cell types.

By transmission electron microscopy, four morphological cell types are distinguishable: lymphocytic coelomocytes (= basophils), granulocytic coelomocytes (= neutrophils and perhaps granulocytes), eleocytes (chloragogen cells), and inclusion-containing coelomocytes (= acidophils). Within these major categories, several distinct cell types differ and may represent developmental stages. Two types of lymphocytic coelomocytes are present, differing primarily in cytoplasmic volume and organelle development. Two types of granulocytic coelomocytes are also present but are probably not developmentally related, differing greatly in shape and content. Eleocytes, derived from chloragogen tissue, contain a variety of granules, inclusions and vacuoles. Inclusion-containing coelomocytes appear as two types which may be immature and mature forms.

The various known functions of coelomocytes have been described in detail but studies focus primarily on lumbricid species. The majority of these functions, particularly those related to coelomocytes (amoebocytes), are either defensive or related to wound healing, while those of eleocytes are usually nutritive or possibly excretory. The coelomocytes of *L. terrestris* have figured prominently in studies of the phylogenesis of cell-mediated immunity, including graft rejection, mitogen responses, and certain surface properties including receptors for mitogens (Con A) and SRBC.

A number of major research areas still require exploration, such as cells of the blood vascular system, both in terms of morphology and function, and their relationship to coelomocytes. The coelomocytes of oligochaete families other than Lumbricidae should be studied more extensively, again in terms of both structure and function. For instance, what is the role of granular substances secreted by acidophils and granulocytes of *L. terrestris*? What is the nature of the surface receptor of basophils allowing them to bind sheep erythrocytes? Do oligochaete coelomocytes exhibit cytotoxicity like sipunculid coelomocytes (Boiledieu and Valembois, 1977) or vertebrate lymphocytes? These few questions are by no means representative of all that can be asked of the earthworm's immune system. Understanding it should help to clarify more complex mechanisms, and only then will immunobiology have become pervasive.

Acknowledgements

We wish to thank Drs Janice M. Burke, Claude Chapron, Ragnar Fänge, Roberta T. Hess, D. Scott Linthicum, G. O. Poinar, Jr. and Pierre Valembois, who generously provided us with micrographs. We also wish to thank Dr Richard K. Wright for his help during preparation of the manuscript. This research was supported in part by USPHS grants HD 09333-05, and 1 R01 AI-15976-01.

References

Abdel-Fattah, R. F. (1955). *Proc. Egypt. Acad. Sci.* **10**, 36–50.
Aiyer, K. S. P. (1925). *Ann. Mag. Nat. Hist.* **16**, 31–40.
Avel, M. (1959). *In* "Traité de Zoologie" (P. P. Grassé, Ed.), Vol. 5, pp. 224–470. Masson et Cie, Paris.
Bailey, S., Miller, B. J. and Cooper, E. L. (1971). *Immunology* **21**, 81–86.
Bainton, D. F. and Farquhar, M. G. (1966). *J. Cell Biol.* **28**, 277–301.
Barnes, R. D. (1974). *In* "Invertebrate Zoology", 3rd edn., pp. 284–285. W. B. Saunders, Philadelphia.
Baskin, D. G. (1974). *In* "Contemporary Topics in Immunobiology" (E. L. Cooper, Ed.), Vol. 4, pp. 55–64. Plenum Press, New York.
Beddard, F. E. (1890a). *Quart. Jl. Microsc. Sci.* **30**, 421 (Cited in Stephenson, J. (1932), "The Oligochaeta", p. 165. Clarendon Press, Oxford).

Beddard, F. E. (1890b). *In* "Proc. Zool. Soc. Lond.", pp. 52–69.

Bell, A. W. (1943). *Trans. Amer. Microsc. Soc.* **62**, 81–84.

Benham, W. B. (1901). *Quart. Jl. Microsc. Sci.* **44**, 565–590.

Bessis, M. (1972). "Living Blood Cells and their Ultrastructure", Chap. V. Springer, Berlin, Heidelberg and New York.

Boiledieu, D. and Valembois, P. (1977). *Devl. Comp. Immunol.* **1**, 207–216.

Boveri-Boner, Y. (1920). *Inaug. Diss. Jena. Vjachr. Naturf. Ges. Zürich* **65**, 506. (Cited in Stephenson, J. (1930), "The Oligochaeta", p. 50. Clarendon Press, London).

Burke, J. M. (1974a). *Cell Tiss. Res.* **154**, 83–102.

Burke, J. M. (1974b). *J. exp. Zool.* **188**, 49–63.

Burnet, F. M. (1974). *In* "Contemporary Topics in Immunobiology" (E. L. Cooper, Ed.), Vol. 4, pp. 13–23. Plenum Press, New York.

Cameron, G. R. (1932). *J. Pathol. Bacteriol.* **35**, 933–972.

Chapron, C. (1970). *Archs Zool. exp. gén.* **111**, 217–228.

Chateaureynaud-Duprat, P. (1971). *Arch. Zool. exp. gén.* **112**, 97–103.

Chateaureynaud-Duprat, P. and Izoard, F. (1977). *In* "Developmental Immunobiology" (J. B. Solomon and J. D. Horton, Eds), pp. 33–40. Elsevier/North Holland, Amsterdam.

Cheng, T. C. and Cali, A. (1974). *In* "Contemporary Topics in Immunobiology" (E. L. Cooper, Ed.), Vol. 4, pp. 25–35. Plenum Press, New York.

Cohen, S. and Lewis, H. (1949). *J. biol. Chem.* **180**, 79–91.

Cohen, S. and Lewis, H. (1950). *J. biol. Chem.* **184**, 479–484.

Cook, D. G. (1969). *Biol. Bull., Woods Hole* **136**, 9–27.

Cooper, E. L. (1965a). *Amer. Zool.* **5**, 169.

Cooper, E. L. (1965b). *Amer. Zool.* **5**, 233.

Cooper, E. L. (1968). *Transplantation* **6**, 322–337.

Cooper, E. L. (1969a). *J. exp. Zool.* **171**, 69–73.

Cooper, E. L. (1969b). *Science* **166**, 1414–1415.

Cooper, E. L. (1973a). *In* "Proc. III International Colloquium on Invertebrate Tissue Culture" (J. Rehacek, D. Blaskovic and W. F. Hink, Eds), pp. 381–404. Publishing House of the Slovak Academy of Sciences, Bratislava.

Cooper, E. L. (1973b). *In* "Symposium on Non-Specific Factors Influencing Host Resistance" (W. Braun and J. S. Ungar, Eds), pp. 11–23. Karger, Basel.

Cooper, E. L. (1976a). *In* "Phylogeny of Thymus and Bone Marrow-Bursa Cells" (R. K. Wright and E. L. Cooper, Eds), pp. 9–18. Elsevier/North Holland, Amsterdam.

Cooper, E. L. (1976b). *In* "Comparative Immunology" (J. J. Marchalonis, Ed.), pp. 36–79. Blackwell Scientific, Oxford.

Cooper, E. L. (1976c). "Comparative Immunology", Prentice Hall, Englewood Cliffs, New Jersey.

Cooper, E. L. (1977). *In* "Phylogenetic Aspects of Transplantation" (J. W. Masshoff, Ed.), pp. 139–167. Springer Verlag, Berlin.

Cooper, E. L. and Rubilotta, L. (1969). *Transplantation* **8**, 220–223.

Cooper, E. L., MacDonald, H. R. and Sordat, B. (1980). *In* "Lymphatic Tissues and Germinal Centers" (W. Müller-Rucholtz and H. K. Müller-Hermelink, Eds), in press.

Cuénot, L. (1891). *Archs Zool. exp. gén.* **9**, 613–641.

Cuénot, L. (1898). *Archs Biol.* **15**, 79–124.

Dales, R. P. (1967). "Annelids". Hutchinson and Co., Ltd., London.

DeBock, M. (1900). *Rev. Suisse Zool.* **8** (Cited in Stephenson, J. (1930), "The Oligochaeta", pp. 168–169. Clarendon Press, London).

Duprat, P. (1964). *C.R. Acad. Sci. Ser. D* **259**, 4177–4180.

Duprat, P. (1967). *Ann. Inst. Pasteur, Paris* **113**, 867–881.

Duprat, P. (1970). *Transplant. Proc.* **3**, 222–225.

Duprat, P. and Bouc-Lassalle, A. M. (1967). *Bull. Soc. Zool. Fr.* **92**, 767–778.

Edwards, C. A. and Lofty, J. R. (1977). "Biology of Earthworms". Chapman Hall, London.

Eisen, G. (1895). *Mem. Calif. Acad. Sci.* **2** (Cited in Stephenson, J. (1930), "The Oligochaeta", p. 165. Clarendon Press, Oxford).

Eisen, G. (1900). *Proc. Calif. Acad. Sci. Ser.* **3D 2**, 85–276.

Fänge, R. (1950). *Ark. Zool.* **1**, 259–264.

Fänge, R. (1974). *In* "Recherches Biologiques Contemporaires", pp. 13–24. Vagher, Nancy.

Fischer, E. (1972). *Acta Histochem.* **42**, 10–14.

Fischer, E. (1975). *Acta Biol. Acad. Sci. Hung.* **26**, 135–140.

Fischer, E. (1977). *Exp. Geront.* **12**, 69–74.

Fischer, E. and Horvath, I. (1978). *Histochemistry* **56**, 165–171.

Freudwiler, H. (1905). *Jena Z. Naturwiss.* **39** (Cited in Semal Van-Gansen (1956). *Bull. biol. Fr. Belg.* **90**, 335–356).

Goodrich, E. S. (1897). *Quart. Jl. Microsc. Sci.* **39**, 51–68.

Grant, W. C. (1955). *Proc. U.S. Nat. Mus.* **105**, 49–63.

Grégoire, C. and Tagnon, H. J. (1962). *In* "Comparative Biochemistry", Vol. IV, pp. 435–482. Academic Press, New York.

Gungl, O. (1905). *Arb. Zool. Inst. Univ. Wien* **15**, 155 (Cited in Stephenson, J. (1930), "The Oligochaeta", p. 169. Clarendon Press, Oxford).

Hama, K. (1960). *J. Biophys. Biochem. Cytol.* **7**, 717–723.

Hanson, J. (1949). *Biol. Rev. Cambridge Philos. Soc.* **24**, 127–173.

Heidermanns, C. (1937). *Zool. Jahrb.* **58**, 57–68.

Hess, R. T. (1970). *J. Morphol.* **132**, 335–351.

Hostetter, R. K. and Cooper, E. L. (1972). *Immunol. Commun.* **1**, 155–183.

Hostetter, R. K. and Cooper, E. L. (1973). *Cell. Immunol.* **9**, 384–392.

Hostetter, R. K. and Cooper, E. L. (1974). *In* "Contemporary Topics in Immunobiology. Invertebrate Immunology" (E. L. Cooper, Ed.), Vol. 4, pp. 91–107. Plenum Press, New York.

Ireland, M. P. and Richards, K. S. (1977). *Histochemistry* **51**, 153–166.

Issel, R. (1906). *Arch. Zool. Ital.* **2**, 125–135.

Keng, L. B. (1895). *Philos. Trans. R. Soc. London, Ser. B* **186**, 383–400.

Kindred, J. E. (1929). *J. Morphol. Physiol.* **47**, 435–468.

Kollmann, M. (1908). *Ann. Sci. Nat. Zool.* **8**, 1–240.

Kükenthal, W. (1885). *Jena Z. Naturwiss.* **18**, 319–355.

Lemmi, C. (1975a). "Tissue graft rejection mechanisms in the earthworm *Lumbricus terrestris*: specific induction of coelomocyte proliferation". Ph.D. thesis. Univ. of Calif., Los Angeles. University Microfilms, Inc., Ann Arbor, Michigan.

Lemmi, C. A. E. (1975b). *Anat. Rec.* **181**, 409.

Lemmi, C. A., Cooper, E. L. and Moore, T. C. (1974). *In* "Contemporary Topics in Immunobiology" (E. L. Cooper, Ed.), Vol. 4, pp. 110–120. Plenum Press, New York.

Liebmann, E. (1926). *Zool. Anz.* **69**, 65.

Liebmann, E. (1927). *Zool. Jahrb. Abt. Allg. Zool. Physiol. Tiere.* **44**, 269–286.

Liebmann, E. (1931). *Zool. Jahrb. Abt. Anat. Ontog. Tiere.* **54**, 417–434.
Liebmann, E. (1942). *J. Morphol.* **71**, 221–249.
Liebmann, E. (1943). *J. Morphol.* **73**, 583–610.
Lindner, E. (1965). *Z. Zellforsch. mikrosk. Anat.* **66**, 891–913.
Ling, N. R. (1968). "Lymphocyte Stimulation". North Holland, Amsterdam.
Linthicum, D. S. (1975). *In* "Immunologic Phylogeny" (W. H. Hildemann and A. A. Benedict, Eds), *Adv. Exp. Med. Biol.* Vol. 64, pp. 241–250. Plenum Press, New York.
Linthicum, D. S., Stein, E. A., Marks, D. H. and Cooper, E. L. (1977a). *Cell Tiss. Res.* **185**, 315–330.
Linthicum, D. S., Marks, D. H., Stein, E. A. and Cooper, E. L. (1977b). *Eur. J. Immunol.* **7**, 871–876.
Lison, L. (1928). *Archs Biol.* **38**, 411–455.
Marks, D. H. and Cooper, E. L. (1977). *J. Invertebr. Pathol.* **29**, 382–383.
Marks, D. H., Stein, E. A. and Cooper, E. L. (1979). *Devl. Comp. Immunol.* **3**, 277–285.
Matsumoto, M. (1960). *Sci. Rep. Tôhuku Univ., Ser. 4*, **26**, 97–105.
Meglitsch, P. A. (1967). "Invertebrate Zoology". Oxford University Press, New York.
Metchnikoff, E. (1891). "Lectures on the Comparative Pathology of Inflammation". (English transl. F. A. and E. H. Starling, 1968), pp. 67–73. Dover Press, New York.
Möller, G. (Ed.) (1972). "Lymphocyte Activation by Mitogens: Models for Immunocyte Triggering". *Transplant. Rev.*, Vol. 11. Williams and Wilkins, Baltimore, Maryland.
Moment, G. B. (1974). *Growth* **38**, 209–218.
Needham, A. E. (1957). *J. exp. Biol.* **34**, 425–446.
Nielsen, C. O. and Christensen, B. (1959). *Nat. Jutl.* **8** and **9**, 1–160.
Nielsen, C. O. and Christensen, B. (1961). *Nat. Jutl.*, suppl. 1 and 2, **10**, 1–23; 1–19.
Nusbaum, J. (1902). *Biol. Zentralbl.* **22**, 292–298.
Ohuye, T. (1934). *Sci. Rep. Tôhuku Univ., Ser. 4*, **9**, 53–59.
Ohuye, T. (1937). *Sci. Rep. Tôhuku Univ., Ser. 4*, **12**, 255–263.
Oppenheim, J. J. and Rosenstreich, D. L. (Eds) (1976). "Mitogens in Immunobiology". Academic Press, New York.
Parry, M. J. (1978). *J. Invertebr. Pathol.* **31**, 383–388.
Poinar, G. O., Jr. and Hess, R. T. (1977). *In* "Comparative Pathobiology" (L. A. Bulla and T. C. Cheng, Eds), Vol. 3, pp. 69–84. Plenum Press, New York.
Prentø, P. (1979). *Cell Tiss. Res.* **196**, 123–134.
Roch, P. (1977). *C.R. Acad. Sci. Ser. D* **284**, 705–708.
Roch, P. and Valembois, P. (1978). *Devl. Comp. Immunol.* **2**, 51–63.
Roch, P., Valembois, P. and Du Pasquier, L. (1975). *Adv. exp. Med. Biol.* **64**, 45–54.
Roots, B. I. (1960). *Comp. Biochem. Physiol.* **1**, 218–226.
Roots, B. I. and Johnston, P. V. (1966). *Comp. Biochem. Physiol.* **17**, 285–288.
Rosa, D. (1896). *Archs Ital. Biol.* **25**, 455–458.
Schneider, G. (1896). *Z. Wiss. Zool.* **61**, 363–392.
Semal Van-Gansen, P. (1956). *Bull. biol. Fr. Belg.* **90**, 335–356.
Semal Van-Gansen, P. (1958). *Enzymologia* **20**, 98–108.
Spencer, W. B. (1888). *Trans. R. Soc. Vict.* **1** (Cited in Stephenson, J. (1930), "The Oligochaeta". Clarendon Press, London).
Sperber, C. (1948). *Zool. Bidr. Uppsala* **28**, 1–296.

Stang-Voss, C. (1971). *Z. Zellforsch. mikrosk. Anat.* **117**, 451–462.

Stang-Voss, C. (1974). *In* "Contemporary Topics in Immunobiology. Invertebrate Immunology" (E. L. Cooper, Ed.), Vol. 4, pp. 65–76. Plenum Press, New York.

Stein, E. A. and Cooper, E. L. (1978). *Histochem. J.* **10**, 657–678.

Stein, E. A., Avtalion, R. R. and Cooper, E. L. (1977). *J. Morphol.* **153**, 467–477.

Stephenson, J. (1924). *Proc. R. Soc. London, Ser. B* **97**, 177–209.

Stephenson, J. (1930). "The Oligochaeta". Clarendon Press, Oxford.

Thapar, G. S. (1918). *Rec. Indian Mus.* **15**, 69–76.

Toupin, J. and Lamoureux, G. (1976a). *Cell. Immunol.* **26**, 127–132.

Toupin, J. and Lamoureux, G. (1976b), *In* "Phylogeny of Thymus and Bone Marrow-Bursa Cells" (R. K. Wright and E. L. Cooper, Eds), pp. 19–26. North Holland, Amsterdam.

Toupin, J., Leyva, F. and Lamoureux, G. (1977). *Ann. Immunol.* (*Paris*) **128c**, 29.

Urich, K. (1960). *Zool. Beitr.* **5**, 281–289.

Valembois, P. (1963). *C.R. Acad. Sci. Ser. D* **257**, 3227–3228.

Valembois, P. (1968). *J. Microsc.* **7**, 61a.

Valembois, P. (1970). "Etude d'une hétérograffe de paroi du corps chez les lombriciens". Ph.D. thesis, University of Bordeaux, Talence, France.

Valembois, P. (1971a). *Arch. Zool. exp. gén.* **112**, 97–104.

Valembois, P. (1971b). *Bull. Soc. Zool. Fr.* **96**, 59–72.

Valembois, P. (1971c). *C.R. Acad. Sci. Ser. D* **272**, 2097–2130.

Valembois, P. (1974). *In* "Contemporary Topics in Immunobiology" (E. L. Cooper, Ed.), Vol. 4, pp. 121–126. Plenum Press, New York.

Valembois, P. and Roch, P. (1977). *Biol. Cell.* **28**, 81–82.

Varute, A. T. and More, N. K. (1972). *Acta Histochem.* **44**, 144–151.

Varute, A. T. and More, N. K. (1973). *Comp. Biochem. Physiol.* **45A**, 607–635.

Vejdovsky, R. (1884). "System und Morphologie der Oligochaeten". Prag. (Cited in Stephenson, J. (1930). "The Oligochaeta", pp. 168–169. Clarendon Press, London).

Vejdovsky, F. (1905). *Z. Wiss. Zool.* **82**, 5–170.

Vejdovsky, F. (1906). *Z. Wiss. Zool.* **85**, 48–73.

Vostal, Z. (1971). *Biologia* (*Bratislava*) **26**, 589–600.

Willem, V. and Minne, A. (1900). *Mem. Acad. R. Belg. Cl. Sci.* **58**, 1–73.

5. Hirudineans

R. T. SAWYER[1] AND S. W. FITZGERALD[2]

[1] *Department of Molecular Biology, University of California, Berkeley, California 94720, U.S.A.* [2] *Department of Zoology, University College of Swansea, Singleton Park, Swansea SA2 8PP, U.K.*

CONTENTS

I. Introduction

The Hirudinea, commonly called leeches or bloodsuckers, is a specialized group of segmented worms closely related to the Oligochaeta (Michaelsen, 1919). It is composed of about 650 species widely distributed throughout the world in diverse freshwater, marine and terrestrial habitats (Harant and Grassé, 1959; Mann, 1961). Apart from the rare, primitive species *Acanthobdella peledina* (Acanthobdellida), true leeches fall naturally into two major orders, those with a proboscis (Rhynchobdellida) and those without a proboscis (Arhynchobdellida) (Table I). In both of these orders, the coelom is reduced to a complicated network of sinuses and this has been brought about by the proliferation of connective tissue. These highly specialized coelomic sinuses form an integral part of the circulatory system, often at the expense of the true vascular system which tends to be reduced (Fig. 1). The arhynchobdellid circulatory system consists entirely of coelomic sinuses and the vascular system is wanting. In this context, the circulatory system of arhynchobdellids can be justifiably described as a haemocoelom and its fluid content haemocoelomic fluid ("blood").

Many aspects of the anatomy and physiology of the circulatory system of leeches have been extensively studied but have not been previously reviewed. The remainder of this chapter describes the gross morphology of the leech circulatory system, the composition of the haemocoelomic fluid ("blood"), and the structure and function of the free "blood" cells and their roles in phagocytosis and tissue repair.

II. Structure of the circulatory system

Leeches are divided into those species with both true vascular and coelomic systems (orders Acanthobdellida and Rhynchobdellida), and those in which the true vascular system has been functionally replaced by coelomic sinuses (order Arhynchobdellida) (Fig. 1).

A. Order Acanthobdellida

Of particular phylogenetic interest is *Acanthobdella peledina* (Table I), whose circulatory system is intermediate between that of the true leeches and oligochaetes (Livanow, 1906). In this annectant species, only the anterior segments have an extensive perivisceral coelom divided into compartments by intersegmental septa similar to those in the Oligochaeta. In all other leeches, such unmodified segmental coeloms are found only in the early embryo (Schleip, 1939).

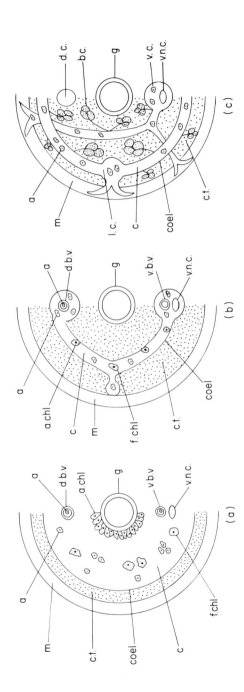

Fig. 1. Phylogenetic scheme showing cross-sections of three stages in the evolution of the chloragogen cell and the haemocoelomic systems. (a) Acanthobdellida. True vascular system and spacious coelom; chloragogen cells. (b) Rhynchobdellida. True vascular system and reduced coelom; attached chloragogen cells also termed "acid cells". (c) Arhynchobdellida. No true vascular system; distinct coelomic channels; botryoidal tissue. Original. a. amoebocyte; b.c. botryoidal cell; c. coelom; a.chl. attached chloragogen cell; f.chl. free chloragogen cell; coel. coelothelium; c.t. connective tissue; d.b.v. dorsal blood vessel; d.c. dorsal channel; g. gut; l.c. lateral channel; m. muscles; v.b.v. ventral blood vessel; v.c. ventral channel; v.n.c. ventral nerve cord.

TABLE I. Taxonomic summary of the species of leeches discussed in the text.

Order	Family	Species	Type of circulatory system	Type of "blood" cells in true vascular system	Type of cells in in coelomic system
Acanthobdellida					
	Acanthobdellidae	Acanthobdellia peledina	vascular and coelomic	amoebocyte attached valvular cell	amoebocyte free chloragogen cell attached chloragogen cell
Rhynchobdellida					
	Piscicolidae	Piscicola geometra Piscicola salmositica	vascular and coelomic	amoebocyte attached valvular cell	amoebocyte attached chloragogen cell (acid cell)[a]
	Glossiphoniidae	Glossiphonia complanata Theromyzon tessulatum Hemiclepsis marginata Placobdella parasitica Placobdella costata Helobdella stagnalis Batracobdella paludosa	vascular and coelomic	amoebocyte attached valvular cell	amoebocyte free chloragogen cell attached chloragogen cell (acid cell)

Arhynchobdellida

Erpobdellidae	Erpobdella octoculata	coelomic only	—	amoebocyte attached chloragogen cell (botryoidal tissue)[b]
Hirudinidae	Hirudo medicinalis Hirudinaria granulosa Haemopis sanguisuga	coelomic only	—	amoebocyte attached chloragogen cell (botryoidal tissue)[b]

[a] Free chloragogen cells probably exist in the Piscicolidae but apparently have not yet been descirbed.
[b] The presence of free chloragogen cells is questionable.

Bearing in mind the close phylogenetic affinities between leeches and oligochaetes, the reader is referred to the preceding chapter which treats in detail the cellular constituents of the coelomic and vascular systems of the Oligochaeta.

B. Order Rhynchobdellida

The vascular system of the rhynchobdellids (Table I), e.g. *Glossiphonia complanata, Hemiclepsis marginata, Theromyzon tessulatum* and *Pisciola geometra* (Whitman, 1878; Bourne, 1884; Oka, 1894, 1902; Livanow, 1910; Selensky, 1907, 1923; Hotz, 1938) consists of a dorsal and a ventral medial vessel, both extending most of the length of the body.

The dorsal vessel carries the blood toward the head and the ventral vessel delivers blood toward the tail. Both vessels lie in a coelomic sinus which functions as an auxiliary circulatory system (Figs 1, 2, 3). Intermediate coelomic channels connect these dorsal and ventral sinuses with a lateral sinus on either side of the body (Fig. 3). A sub-epidermal network continuous with the coelomic system is also present. The blood system is closed and is apparently never in direct contact with the coelomic system.

Leydig (1849) was the first to recognize that the rhynchobdellid circulatory system consists of a closed contractile system and an open non-

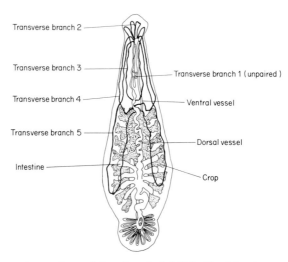

Fig. 2. True vascular system of the rhynchobdellid, *Hemiclepsis marginata* (Oka, 1894). Dorsal view. Not all of the crop cacca are shown.

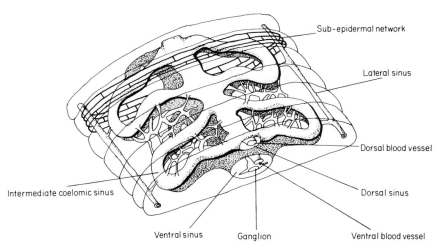

Fig. 3. Relationship of the coelomic system with the true vascular system in the rhynchobdellid, *Theromyzon tessulatum* (Hotz, 1938). Dorsal view.

contractile system (a minor exception is the piscicolid circulatory system whose lateral sinuses are contractile). The closed system of rhynchobdellids is homologous with the true vascular system of oilgochaetes, and the open system of rhynchobdellids corresponds to the oligochaete coelom (Livanow, 1906).

The contractile dorsal blood vessel gives rise anteriorly to four paired transverse branches, each of which describes an individual loop and eventually connects with the ventral blood vessel. An additional unpaired branch serves the proboscis (Fig. 2). In its mid-region the dorsal vessel dilates to form a series of 15–20 chambers which are associated with the cellular valves to the walls of the vessel (Fig. 4) (Hotz, 1938). Each valve

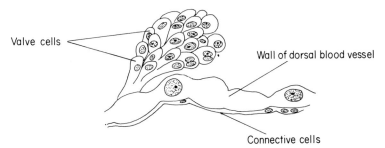

Fig. 4. Valve of the dorsal blood vessel of the rhynchobdellid, *Glossiphonia complanata* (Oka, 1894).

consists of 40–50 cells attached to a stalk which is continuous with the wall of the vessel. Behind each valve is a sphincter consisting of a few small circular muscle cells. The sphincters and valves are presumably important in directing the flow of blood anteriorly (Johansson, 1896; Hotz, 1938). The structure of the non-contractile ventral blood vessel is uniform along its entire length.

C. Order Arhynchobdellida

The true vascular system has completely disappeared in the arhynchobdellids (Table I), in which the coelomic channels function as a haemodynamic system (Fig. 1) (Bourne, 1884). "Blood," or more properly haemocoelomic fluid, circulates through distinct channels derived exclusively from the coelomic system. Our understanding of the arhynchobdellid haemocoelomic system is based primarily on *Hirudo medicinalis* (Bourne, 1884; Jaquet, 1886; Goodrich, 1899; Gaskell, 1914; Boroffka and Hamp, 1969) in which there are four main longitudinal haemocoelomic channels (Fig. 5): dorsal, ventral and paired laterals. The dorsal channel, which corresponds to the rhynchobdellid dorsal coelomic sinus extends mid-dorsally between the head and caudal sucker and gives rise in each segment to two pairs of dorsal cutaneous branches joining the capillary plexus to the skin. The ventral channel lies beneath the digestive tract extending mid-ventrally between the oral and causal suckers and gives rise to two pairs of branches in each segment. It entirely encloses the ventral nerve cord. The ventral channel lacks muscles, valves (Bhatia, 1977) and sphincters (Scriban, 1936). The relatively large lateral channels have acquired muscular walls and are responsible for circulating the haemocoelomic fluid (Hammersen and Staudte, 1969). In each segment they each give rise to three main branches: the latero-ventrals, latero-dorsals and latero-laterals (Fig. 5).

The lateral channels contract more or less alternately at about 6–15 per minute at 21°C depending on the activity of the animal (Hammersen *et al.*, 1976). Boroffka and Hamp (1969) helped elucidate the course of the haemocoelomic fluid in *Hirudo medicinalis* by the injection of dyes. Additional observations have been made on *Hirudinaria granulosa* (Bhatia, 1977). The contraction of the lateral channels is under complex neuronal control which has been thoroughly investigated (Thompson and Stent, 1976). The fluid in the two contractile lateral channels flows anteriorly and that of the non-contractile dorsal and ventral channels flows posteriorly. The direction of fluid is regulated, in part at least, by muscular sphincters in the lateral channels.

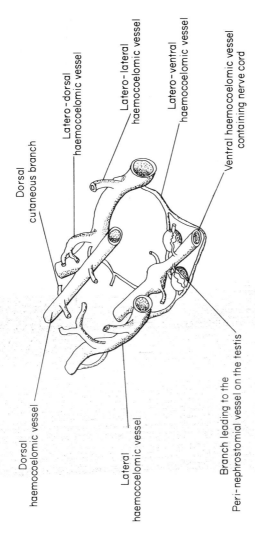

Dorsal
cutaneous branch

Latero-dorsal
haemocoelomic vessel

Latero-lateral
haemocoelomic vessel

Latero-ventral
haemocoelomic vessel

Ventral haemocoelomic vessel
containing nerve cord

Dorsal
haemocoelomic vessel

Lateral
haemocoelomic vessel

Branch leading to the
Peri-nephrostomial vessel on the testis

Fig. 5. Haemocoelomic channels in a mid-body segment of *Hirudo medicinalis* (Mann, 1961).

III. Composition of haemocoelomic fluid

The haemocoelomic fluid or "blood" is a liquid matrix probably secreted by the connective tissue. In *H. medicinalis* the anions of the haemocoelomic fluid include nitrate, phosphate, sulphate as well as organic anions (Nicholls and Kuffler, 1964; Zerbst-Boroffka, 1970). The major known cations are sodium, calcium and potassium. The amino acid content is low (Boroffka, 1968). The localization of exopeptidase and leucylaminopeptidase in the valvular apparatus of arhynchobdellids has led to the proposal that the valve may be involved in the regulation of the chemical composition of the haemocoelomic fluid (Fischer, 1970). Haemoglobin was identified in 1872 as the red pigment of the circulating fluid of arhynchobdellids (Lankester, 1872). The circulating fluid of rhynchobdellids is nearly colourless and the presence of a respiratory pigment has not been well substantiated. A water-soluble yellow pigment has been isolated from three glossiphoniids and a piscicolid (Needham, 1966), the nature of which is not known. In this context the reported presence of haemoglobin in *Placobdella parasitica* requires verification (Shlom *et al.*, 1975).

IV. Structure and classification of "blood" cells

The nature of the free "blood" cells in leeches has received little attention. In line with many other invertebrates, pleomorphism and subcellular diversity account for a rather confused understanding of the origin, function and especially the terminology of leech "blood" cells. Basically, two types of cells occur in the coelomic and haemocoelomic fluids of leeches: (a) amoebocytes and (b) free chloragogen cells.

A. Amoebocytes

Amoebocytes, also called leucocytes (Abeloos, 1925) or lymphocytes (Scriban, 1936), are colourless, amoeboid cells which occur freely or attached to the walls of the haemocoelomic vessels and have a tendency to send out many filamentous pseudopodia (Fig. 6). The cytoplasm contains a diversity of inclusions, as well as vacuoles which seem to change with age (Abeloos, 1925) or with the phagocytic history of the cell. Cell size tends to be rather uniform in most species studied, e.g. *Hemiclepsis marginata* and *Glossiphonia complanata* (Holtz, 1938) where the amoebocytes are 6–7 μm in diameter with nuclei 3–4 μm across. However, in other species they may be noticeably larger, e.g. up to 10 μm in diameter with nuclei 4–6 μm across in

Theromyzon tessulatum (Hotz, 1938). Under the electron microscope, the amoebocytes of *Haemopis sanguisuga* all appear very similar in structure and apparently belong to a single cell lineage with the nuclear/cytoplasmic ratio decreasing as the cells mature (Figs 7–9). The amoebocytes typically have an oval nucleus, containing peripherally arranged heterochromatin (Figs 7, 8), and are surrounded by basophilic cytoplasm. The cytoplasm has a very characteristic structure and is honeycombed by a variable number of invaginations and/or vacuoles which contain vesicular material very similar to that present outside the cells (Fig. 9). This material may well represent haemoglobin which has been ingested from or is being discharged into the "blood". The peripheral cytoplasm is also characterized by bundles of microtubules and often has a few small membrane-bound granules, mitochondria, an occasional Golgi and one or two profiles of endoplasmic reticulum (Figs 7–9).

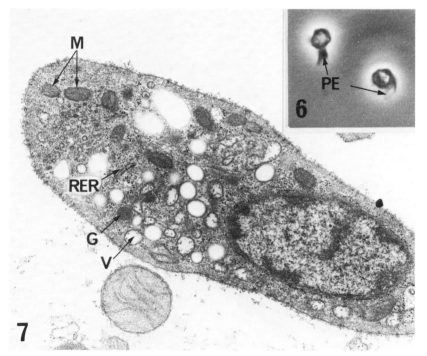

Fig. 6. Amoebocytes of *Haemopis sanguisuga* putting out protoplasmic extensions (PE) after approx. five minutes *in vitro*. × 1400.

Fig. 7. Electron micrograph of an amoebocyte of *H. sanguisuga*. The cytoplasm contains vacuoles (V), mitochondria (M), sparse rough endoplasmic reticulum (RER) and occasional granules (G). × 22 000.

In the Acanthobdellida and Rhynchobdellida the coelomic system is separate from the closed vascular system. Amoebocytes characteristic of the latter appear to be smaller than those of the coelomic system (Scriban, 1907, 1936; Abeloos, 1925; Babaskin, 1931; Scriban and Epure, 1931; Hotz, 1938). Accordingly, some authors believe that amoebocytes of the closed vascular system are distinct from those of the coelomic system (Scriban, 1907; Abeloos, 1925). Available evidence, however, does not appear to support this hypothesis (Oka, 1894; Cuénot, 1897; Livanow, 1906; Babaskin, 1931).

The majority of amoebocytes appear to move freely in the fluid of the haemocoelomic vessels. Others attach to the epithelium of the vessel walls and may even migrate into the connective tissue (Livanow, 1906) where they are most abundant around the gut (Babaskin, 1931) or in the vicinity of an injury (Le Gore and Sparks, 1973). Amoebocytes also aggregate in large numbers within the cilio-phagocytic organ of the nephridia (Fig. 10) (Bourne, 1884) and coelomic sinuses of both the testisacs and ovisacs (Brumpt, 1900).

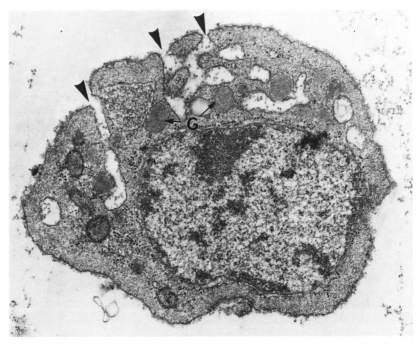

Fig. 8. Electron micrograph of an amoebocyte of *H. sanguisuga* showing the characteristic invaginations of the outer membrane (unlabelled arrows) to form "intracellular canals". Granules (G). × 30 000.

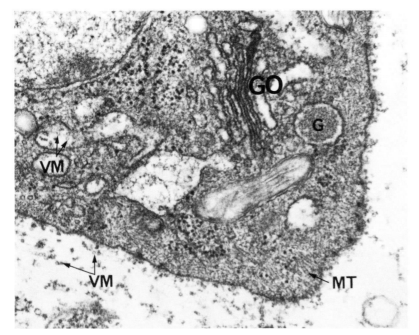

Fig. 9. Part of the cytoplasm of an amoebocyte of *H. sanguisuga* showing a Golgi complex (GO) and associated granule (G). Note that the vacuoles/infoldings contain a vesicular material (VM) similar to that seen in the "blood". Microtubules (MT). × 52 000.

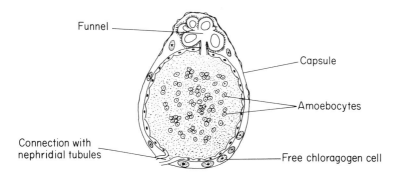

Funnel

Capsule

Amoebocytes

Connection with
nephridial tubules

Free chloragogen cell

Fig. 10. Cilio-phagocytic organ of *Pontobdella muricata*, showing amoebocytes and free chloragogen cells (Bourne, 1884).

B. *Free chloragogen cells*

A second, larger type of cell is less frequently encountered in the "blood". This is the chloragogen cell, which has become detached from the coelomic walls and moves about freely in the coelomic fluid as "coelomocytes" (Fig. 1) (Oka, 1894; Graf, 1899; Livanow, 1906; Scriban, 1910; Abeloos, 1925; Hotz, 1938). They may also be found in the cilio-phagocytic organ of the nephridia (Fig. 10) (Bourne, 1884; Kowalevsky, 1897; Abeloos, 1925; Hotz, 1938).

The size of free chloragogen cells varies with the age and species of the individual. In *Acanthobdella peledina*, they tend to be rounded cells 12–16 µm in diameter with a spherical nucleus of 6–8 µm containing 1 or 2 nucleoli (Livanow, 1906). The attached chloragogen cells of rhynchobdellids (Fig. 1b) were originally called "acid cells" due to their staining charac-teristics (Kowalevsky, 1897). These acid cells are homologous to the chloragogen cells of oligochaetes and botryoidal cells of arhynchobdellids (Bradbury, 1959). In most glossiphoniids, each chloragogen cell is rounded and about 15–25 µm in diameter (Hotz, 1938). They possess an ovoid nucleus of 6–8 µm containing a large nucleolus.

Cytoplasmic inclusions of the free chloragogen cell vary considerably with the nutritional state of the individual (Babaskin, 1931; Hotz, 1938) (Figs 11a,b). Lipid droplets occur in well-nourished individuals (Livanow, 1906; Hotz, 1938). Numerous pigment granules, which may be derived from the breakdown of haemoglobin, occur in the chloragogen cells of those which feed on vertebrate blood (Iuga, 1931; Bradbury, 1959) (Figs 11a,b). Such inclusions are inconspicuous in those species which do not feed on vertebrate blood (Scriban, 1907; Abeloos, 1925).

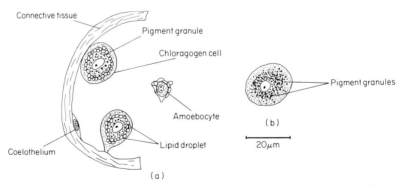

Fig. 11. Chloragogen cells of adult *Theromyzon tessulatum* (Hotz, 1938). (a) Chloragogen cells of a fed individual, attached to the coelomic wall; note the amoebocyte. (b) Free non-vacuolated chloragogen cell of a hungry individual.

V. Origin and formation of "blood" cells

Throughout the Hirudinea the coelomic cavities are lined by thin epithelial cells (coelothelium) (Fig. 11a). Some of these epithelial cells, as discussed above, have been further modified into specialized chloragogen cells (termed botryoidal cells in Arhynchobdellida) and perhaps amoebocytes (Fig. 1). An unequivocal demonstration of the origin of amoebocytes has been elusive and the several proposals put forward to date continue to be speculative. The most widespread view is that leech amoebocytes arise from the coelomic epithelium (Livanow, 1906; Babaskin, 1931; Hotz, 1938). To what extent free amoebocytes can multiply is unclear, but in *Acanthobdella peledina* multinucleated amoebocytes are occasionally seen in mitosis (Livanow, 1906). Some workers have argued that at least some amoebocytes (those of the closed system of rhynchobdellids) arise from the specialized endothelial cells which make up the valves of the dorsal blood vessel (Oka, 1894; Arnesen, 1904) (Fig. 4). That amoebocytes arise from the nephridial capsule or cilio-phagocytic organ (Fig. 10) (Abeloos, 1925; Bhatia, 1938) seems unlikely.

VI. Functions of "blood" cells

"Blood" cells in leeches phagocytose foreign bodies, and they also transport waste products and food, and take part in wound healing.

A. Phagocytosis

Bacteria or inert particles such as carmine and India ink are phagocytosed by amoebocytes. Chloragogen cells take up injected carmine but fail to ingest other substances, e.g. bacteria or India ink (Kowalevsky, 1897; Willem and Minne, 1899; Abeloos, 1925; Iuga, 1931). Rhynchobdellid amoebocytes will also phagocytose some of the spermatozoa which normally enter the coelom in channels following mating (Brumpt, 1900; Abeloos, 1925). Phagocytic activity by the leech amoebocyte has frequently been observed but its subsequent fate is not clear. Emden (1929) claimed that in *Erpobdella octoculata*, amoebocytes migrate into the lumen of the intestine. This has been observed in the polychaete, *Arenicola marina*, in which carbon particles ingested by coelomocytes have also been found to migrate through the intestinal wall (Kermack, 1955) (see Chapter 3). In glossiphoniid leeches, amoebocytes containing ingested bacteria or carmine are concentrated in the cilio-phagocytic organ of the nephridia (Kowalevsky, 1897). Their ultimate fate once they have arrived there is again unclear. Particulate matter in rhynchobdellids is taken up by the cells of the coelomic epithelium (Tilloy, 1937) and in arhynchobdellids by specialized groups of cells constituting the vaso-fibrous (=specialized connective tissue) and botryoidal tissue (Bradbury, 1959).

It may be the case, as in other invertebrates, that much waste material as well as foreign matter accumulate in the epidermis and other tissues and remain within the body indefinitely.

B. Wound healing

Superficial wounding in leeches leads to the formation of a permanent scar composed of poorly differentiated connective tissue (Würgler, 1918; Myers, 1935; Le Gore and Sparks, 1971, 1973; Cornec and Coulomb-Gay, 1975). The role of amoebocytes in this process is not fully understood. Immediately following traumatization muscles in the vicinity of the wound contract to seal the wound, thus minimizing the loss of body fluids. Shortly afterwards, cells of most tissue types undergo apparent dedifferentiation. Numerous undifferentiated cells of undetermined origin aggregate in the subepithelial region to form a pseudoblastema (Myers, 1935). The latter is composed of two predominant cell types distinguished by the nature of their nuclei rather than by cell size: (1) very numerous fibrocyte-like cells with cytoplasmic extensions and nuclei (2·5–4·0 μm in *Placobdella parasitica*) and (2) less numerous amoeboid cells with inconspicuous cell membranes and larger nuclei (5–7 μm in *P. parasitica*). The amoeboid cells may consist of cells dedifferentiated *in situ* and/or amoebocytes migrating from elsewhere in the body. Other cell types such as the free chloragogen cells are only occasionally found in the pseudoblastema. Severely damaged tissue is sloughed off to the exterior or attacked by amoebocytes (Myers, 1935; Le Gore and Sparks, 1971, 1973). The pseudoblastema fails to differentiate further. The only apparent exception to this is the pseudoblastema resulting from histolysis due to hypodermic implantation of a spermatophore, e.g. in *P. parasitica* (Myers, 1935); in this case the blastema supposedly differentiates into nearly normal tissue. A new epithelium, formed primarily by an overgrowth of healthy epithelial cells from the periphery of the wound, covers the pseudoblastema. The resultant, permanent scar is discoloured due to the lack of pigment and gland cells in the new epithelium (Hirschler, 1907; Myers, 1935; Cornec, 1971; Le Gore and Sparks, 1971).

The repair process in leeches is far inferior to that of oligochaetes and is probably related to the inability of leeches to regenerate lost segments or organs (Cornec and Coulomb-Gay, 1975).

VII. Summary and concluding remarks

In the present chapter, the gross morphology of the circulatory system, which has been extensively studied in a number of leeches, is described and reviewed.

Research into the nature of the circulating "blood" cells in leeches has received little attention within the last thirty years, and our understanding of the origins, functions and terminology of leech "blood" cells is confused. Although a number of regions have been suggested an unequivocal demonstration of the site of origin of amoebocytes has been elusive. To what extent free amoebocytes have the capacity to multiply is unclear. It seems likely that leeches lack special, well-defined leucocytopoietic organs and that amoebocytes arise from the coelomic epithelium.

A brief, electron microscopic description of the amoebocytes of *Haemopis sanguisuga* is presented and it appears that a single cell lineage exists. However, a comparative and comprehensive ultrastructural cytochemical study of leech "blood" cells is completely lacking.

The functional capabilities of the amoebocytes are also poorly understood. Phagocytic activity has frequently been observed but the subsequent fate of ingested material is not clear. It is also unknown whether leech amoebocytes are as active as oligochaete coelomocytes in the repair of damaged tissue. Clearly, in recent years research into the cellular defence reactions of leeches has been neglected and should be re-examined especially in the light of the interesting studies which have been carried out on the polychaetes and oligochaetes (see Chapters 3 and 4).

Acknowledgements

We would like to thank Mr Alan Osborn, Drs Norman A. Ratcliffe and Andrew F. Rowley, from the University College of Swansea, for their assistance in the preparation of this paper.

References

Abeloos, M. (1925). *Bull. biol. Fr. Belg.* **59**, 436–497.
Arnesen, E. (1904). *Jena Z. Naturw.* **38**, 771–806.
Babaskin, A. W. (1931). *Zool. Jb. Anat.* **53**, 1–102.
Bhatia, M. L. (1938). *Quart. Jl. Microsc. Sci.* **81**, 27–80.
Bhatia, M. L. (1977). *In* "Memoirs on Indian Animal Types". Emkay Publications, Delhi.
Boroffka, I. (1968). *Z. vergl. Physiol.* **57**, 348–375.

Boroffka, I. and Hamp, R. (1969). *Z. Morph. Tiere* **64**, 59–76.

Bourne, A. G. (1884). *Quart. Jl. Microsc. Sci.* **24**, 419–506.

Bradbury, S. (1959). *Quart. Jl. Microsc. Sci.* **100**, 483–498.

Brumpt, É. (1900). *Mém. Soc. zool. Fr.* **13**, 286–430.

Cornec, J. P. (1971). *Ann. Embryol. Morph.* **4**, 269–279.

Cornec, J. P. and Coulomb-Gay, R. (1975). *C.R. Seances Soc. Biol.* **169**, 86–93.

Cuénot, L. (1897). *Archs Anat. microsc. Paris* **1**, 153–192.

Emden, M. van (1929). *Z. wiss. Zool.* **134**, 1–83.

Fischer, E. (1970). *Acta Histochem.* **37**, 170–175.

Gaskell, J. F. (1914). *Phil. Trans. R. Soc. Lond.* (*B*) **205**, 153–211.

Goodrich, E. S. (1899). *Quart. Jl. Microsc. Sci.* **42**, 477–495.

Graf, A. (1899). *Acta Ac. Leop.* (*Halle*) **72**, 215–404.

Hammersen, F. and Staudte, H. W. (1969). *Z. Zellforsch. mikrosk. Anat.* **100**, 215–250.

Hammersen, F., Staudte, H. W. and Moehring, E. (1976). *Cell Tiss. Res.* 173, 405–424.

Harant, H. and Grassé, P. (1959). *In* "Traité de Zoologie" (P. Grassé, Ed.), Vol. 5, pp. 471–593. Masson, Paris.

Hirschler, J. (1907). *Zool. Anz.* **32**, 212–216.

Hotz, H. (1938). *Revue suisse Zool.* **45**, 1–380.

Iuga, V. G. (1931). *Archs Zool. exp. gén.* **71**, 1–97.

Jaquet, M. (1886). *Mitt. zool. Stn. Neapol.* **6**, 297–398.

Johansson, L. (1896). *Festschr. fur Lilljeborg, Uppsala*, pp. 317–330.

Kermack, D. M. (1955). *Proc. zool. Soc. Lond.* **125**, 347–385.

Kowalevsky, A. (1897). *Mém. Acad. Sci. St. Péterb. phys-math* **5**, 1–15.

Lankester, E. R. (1872). *Proc. R. Soc. Lond.* **21**, 70–81.

Le Gore, R. S. and Sparks, A. K. (1971). *J. Invert. Pathol.* **18**, 40–45.

Le Gore, R. S. and Sparks, A. K. (1973). *J. Invert. Pathol.* **22**, 298–299.

Leydig, F. (1849). *Ber. zootom. Anst. Würzburg* **2**, 14–20.

Livanow, N. (1906). *Zool. Jb. Anat.* **22**, 637–866.

Livanow, N. (1910). *Biol. Z. Moskau* **1**, 46–67.

Mann, K. H. (1961). "Leeches (Hirudinea). Their Structure, Physiology, Ecology and Embryology". Pergamon Press, London.

Michaelson, W. (1919). *Mitt. Naturh. Mus. Hamburg* **36**, 131–153.

Myers, R. J. (1935). *J. Morph.* **57**, 617–648.

Needham, A. E. (1966). *Comp. Biochem. Physiol.* **18**, 427–461.

Nicholls, J. G. and Kuffler, S. W. (1964). *J. Neurophysiol.* **27**, 645–673.

Oka, A. (1894). *Z. wiss. Zool.* **58**, 79–151.

Oka, A. (1902). *Annot. Zool. Jap.* **4**, 49–60.

Schleip, W. (1939). *In* "Klassen und Ordnungen des Tierreichs" (H. G. Bronn, Ed.), Vol. 4, Div. 3 Book 4, Part 2, pp. 1–121. Akademische Verlagsgesellschaft, Leipzig.

Scriban, J. A. (1907). *Archs Zool. exp gén.* **7**, 397–421.

Scriban, J. A. (1910). *Annls. Scient. Univ. Jassy* **6**, 147–286.

Scriban, J. A. (1936). *Zool. Anz.* **113**, 64–67.

Scriban, J. A. and Epure, E. (1931). *Zool. Anz.* **94**, 322–328.

Selensky, W. D. (1907). *Trav. Soc. Natural. St. Pétersb.* **36**, 37–111.

Selensky, W. D. (1923). *Zool. Jb. Syst.* **46**, 397–488.

Shlom, J. M., Amesse, L. and Vinogradov, S. N. (1975). *Comp. Biochem. Physiol. B* **51**, 389–392.

Thompson, W. J. and Stent, G. S. (1976). *J. Comp. Physiol. A*, **111**, 261–279, 281–307, 309–333.

Tilloy, R. (1937). *Bull. mem. Soc. Sci. Nancy (n.s.)* **7**, 199–225.

Whitman, C. O. (1878). *Quart. Jl. Microsc. Sci.* **18**, 215–315.

Willem, V. and Minne, A. (1899). *Mem. Ac. Sci. Belg.* **58**, 51–87.

Würgler, E. (1918). *Viert. Jb. Naturf. Ges. Zürich* **3**, 552–565.

Zerbst-Boroffka, I. (1970). *Z. vergl. Physiol.* **70**, 313–321.

6. Sipunculans and echiuroids

L. DYBAS

Department of Biology, Knox College, Galesburg, Illinois 61401, U.S.A.

CONTENTS

* With the exception of Fig. 14, all the figures in this chapter are from the sipunculan, *Phascolosoma agassizii* Keferstein, found along the coastline of California, U.S.A. Fig. 14 is from *Golfingia gouldii*, Pourtales, collected from the intertidal zone along the coastline of Massachusetts, U.S.A.

PART 1: SIPUNCULANS

I. Introduction

The phylum Sipuncula contains over 300 described species of marine, worm-like animals (Stephen and Edmonds, 1972). They occur in all oceans of the world, primarily in the intertidal zone where they burrow in the substrata or in rock crevices and mussel holdfasts. Some species have been reported at water depths of more than 5000 metres. Sipunculans vary in size, but all possess the same basic body form (Fig. 1) consisting of two major regions: a thick trunk, and a more slender introvert with a mouth at the end surrounded by tentacles and hooks. A muscular body encloses an undivided coelomic cavity lines with a peritoneum of flattened cells which may bear cilia. The internal organs are few: a nerve cord with brain, one or more nephridia, an oesophagus, and a looped, coiled intestine. The organs are attached to the body wall by fixing fibres (Hyman, 1959).

With one exception, sexes are separate (Åkesson, 1958) and there are no apparent external differentiating characteristics. Depending on the sex of the animal, either ova or spermatids are shed from the gonads at the base of the retractor muscles into the coelomic cavity. There they undergo development freely mixing with the circulating coelomocytes. Mature gametes are released through the nephridia and fertilization is external. The embryos undergo spiral cleavage and become trochophore larvae. Asexual reproduction has also been reported in two species of sipunculans (Rajulu and Krishnan, 1969; Rice, 1970). No segmentation has been reported in either larval or adult sipunculans. This is the main criterion for separation of sipunculans from the phylum Annelida, the segmented worms, which are typically organized into metameric subunits.

II. Structure of the circulatory system

Sipunculans do not have a true circulatory system. The coelomocytes and gametes circulate freely in the non-partitioned coelomic cavity which runs from the anterior oesophageal region in the introvert to the posterior end of the animal. Muscular contractions of the body maintain the movement of the coelomic fluid (Hyman, 1959). In some species, this movement is aided by cilia on the peritoneal and freely circulating cells.

III. Origin and formation of coelomocytes

Rice (1973), in her study of the developmental biology of three species of sipunculans, noted two circulating cell types in the larvae: a spindle-shaped cell with granules, and a round, flat cell that over a two month period gradually accumulated granules and vacuoles. These cells appeared to bud from the splanchnic peritoneum into the coelom. Other cell types seen in adult sipunculans have not been observed in the larvae.

No true haemopoietic organs have been described in adult sipunculans, although some researchers have reported that the circulating cells appear to originate from the peritoneal lining of the coelomic cavity or the Polian canal (Hérubel, 1908; Volkonsky, 1933; Ohuye, 1937). Adult cell types have also been seen in mitosis (Thomas, 1931; Thomas, 1932–33; Volkonsky, 1933; Yeager et al., 1935; Ohuye et al., 1961). In some species, freely circulating ciliated urns are thought to arise from fixed urns lining the peritoneal cavity (Selensky, 1908). Immature cells that may represent an undifferentiated stem cell (Dybas, 1973) have been described in *Phascolosoma agassizii*, but the origin of these cells is unknown.

IV. Structure and classification of coelomocytes

The coelomocytes described for the Sipuncula fall into five categories: haemocytes (an unfortunate name for a coelomocyte cell type), granulocytes, large multinuclear cells, ciliated urns and immature cells. Not all cell types are represented in every species. The literature relating to classification of sipunculan coelomocytes is summarized in Table I.

Early attempts to classify and identify coelomocytes were based on light microscope observations aided by use of vital and basic histochemical stains. Results of these studies are reviewed here for comparison in Table II. Histochemical studies which localize specific substances within the cells are also included. Identification of coelomocyte types by basic histological

TABLE I. Classification of Sipuncula coelomocytes.

Genus, species	Haemocytes	Immature cells	Granulocytes
Phascolosoma gouldii	coloured corpuscles (6–24 μm × 3 μm)	—	granular amoebocytes, white corpuscles[a]
Phascolosoma vulgare (*Golfingia*)	hématies or red blood cells (16 μm)	—	amoebocytes[a] young and old acidophils[a] neutrophils (large and small)[a] basophils (large and small)[a]
Sipunculus nudus *Golfingia vulgare* *Phascolosoma elongatum* *Physcosoma granulatum*	hématies[a]	—	amoebocytes, young and old[a]
Sipunculus nudus	pink corpuscles[a]	—	
Sipunculus nudus	erythrocytes[a]	—	leukocytes[a]
Sipunculus nudus	hématies (3 μm)	—	leukocytes[a]
Phascolosoma vulgare			large and small granulocytes[a] hyaline lymphocytes[a] acidophilic spherules[a] basophilic spherules[a]
Physcosoma scolops	erythrocytes (8–20 μm × 2–4 μm)	—	leukocytes[a] hyaline amoebocytes (4–16 μm) acidophilic granulocytes (10–16 μm) basophilic granulocytes (10–16 μm)
Survey of phylum	red corpuscles (6–32 μm)	—	hyaline amoebocytes[a] granular amoebocytes or granulocytes[a]
Phascolosoma agassizii	nucleated red blood cells (14·4 μm)	—	granular phagocytes (16·7 μm diameter) granular amoebocytes (58 μm × 50 μm)
Phascolosoma agassizii	red cells or spherules (6–32 μm)	—	basophils (4·5 μm) acidophils (6·8 μm × 5·8 μm) small granulocytes (16 μm × 8 μm)
Golfingia capensis	haemocytes[a]	—	hyaline amoebocytes[a] mature amoebocytes[a] granulocytes[a]
Golfingia gouldii	haemocytes (30 μm)	prohaemocytes (10 μm)	leukocytes[a]
Phascolosoma agassizii	haemocytes (22 μm × 17·5 μm)	blast cells (10–18 μm)	acidophils (10·9–15·3 μm) basophils (10·9–15·3 μm)
Golfingia ikedai *Golfingia nigra* *Phascolosoma scolops* *Siphonosoma cumanence*	erythrocytes (10–20 μm)	—	leukocytes, small basophilic granules[a] leukocytes, large basophilic granules[a] leukocytes, fine granules[a] phagocytes[a]

[a] Size not measured—no data available.

Large multinuclear cells	Ciliated urns	Researcher
giant corpuscles (32–123 μm × 42 μm)	chloragogues, ciliated cells[a]	Andrews (1890)
vésicles énigmatiques (540 μm)	ciliated cells, fixed urns[a]	Cuénot (1900)
vésicles énigmatiques	motile and fixed urns (52 μm × 38 μm × 17 μm)	Hérubel (1908)
mulberry spheres[a] vésicles énigmatiques (100 μm) vésicles énigmatiques[a]	topfchen (urns) or ciliated globes[a] urns[a] —	Lankester (1908) Thomas (1931, 1932a,b, 1932–33) Volkonsky (1933)
acidophilic granulocytes (16–38 μm × 5–8 μm or 12–30 μm × 5–8 μm)	urns (15–40 μm)	Ohuye (1937)
giant multinuclear corpuscles[a]	urns (fixed or free)[a]	Hyman (1959)
—	free urns (59 μm × 52 μm)	Towle (1962, 1975)
multicellular granulocytes[a]	urns (multicellular) (25 μm)	Blitz (1965)
multinuclear granulocytes[a] vésicles énigmatiques[a]	no free urns	Brown and Winterbottom (1969)
—	—	Stang-Voss (1970)
"cell within a cell" (26–52 μm × 15–30 μm)	free urns (multicellular) (57 μm × 61 μm)	Dybas (1973, 1975, 1976)
morula cells[a] vesicular cells (30–60 μm)	—	Ochi (1975a)

TABLE II. Staining properties of Sipuncula coelomocytes.

Genus, species	Stain	Staining reaction in haemocytes
Sipunculus nudus	Iron lugol	—
	Nile blue sulphate	—
	Sudan III	—
	Osmic acid	—
Phascolosoma gouldii	Wright's neutral red	basophilic granules + vacuoles −
	Janus green B	mitochondria +
	Brilliant cresyl blue	granules, dark blue vacuoles, pink
Phascolosoma vulgare	Silver impregnation	vacuole contents +
Sipunculus nudus	Osmic acid	vacuole contents +
	Brilliant cresyl blue	—
	Neutral red	vacuole contents +
	Toluidine blue	—
Physcosoma scolops	Janus green	mitochondria +
	Giemsa	eosinophilic to basophilic
	Janus green B	mitochondria +
	Neutral red	perinuclear granules +
	Methylene blue	perinuclear granules +
	Nile blue sulphate	perinuclear granules +
	Brilliant cresyl blue	perinuclear granules +
	Osmic acid	perinuclear granules +
	Sudan III	perinuclear granules +
	Scharlach R	—

Staining reaction in granulocytes	Staining reaction in large multinuclear cells	Staining reaction in ciliated urns	Researcher
—	cytoplasm +	—	Thomas (1932–33)
—	cytoplasm +	—	
—	cytoplasm +	—	
—	cytoplasm +	—	
—	—	—	Dawson (1933)
—	—	—	
—	—	—	
—	—	—	Volkonsky (1933)
—	—	—	
acidophilic spherules, sky blue basophilic spherules, colourless small granulocytes, sky blue large granulocytes, colourless leukocytes, violet	—	—	
acidophilic spherules, pinkish purple basophilic spherules, colourless small granulocytes, colourless large granulocytes, colourless leukocytes, red hyaline lymphocytes, red precipitate			
acidophilic spherules, sky blue basophilic spherules, colourless small granulocytes, light blue large granulocytes, colourless leukocytes, violet granules mitochondria +	—	—	
acidophilic granulocytes, eosinophilic granules basophilic granulocytes, basophilic granules	granules, eosinophilic	—	Ohuye (1937)
hyaline amoebocytes, mitochondria + acidophilic granulocytes, mitochondria + basophilic granulocytes, mitochondria +	—	globules +	
acidophilic ¢ basophilic granulocytes +	granules +	globules +	
acidophilic ¢ basophilic granulocytes +	granules +	globules +	
acidophilic ¢ basophilic granulocytes +	granules dissolve	—	
acidophilic ¢ basophilic	—	globules +	
acidophilic ¢ basophilic granulocytes +	—	—	

TABLE II.—*cont.*

Genus, species	Stain	Staining reaction in haemocytes
Phascolosoma scolops	Neutral red	—
	Nile blue sulphate	—
	Victoria blue 4R	—
	Janus green B	—
	(Sterols and compound lipids)[a]	—
Phascolosoma agassizii	Periodic acid Schiff	—
	Haemotoxylin and eosin	cytoplasm, blue
Phascolosoma agassizii	Wright's stain	basophilic
	May-Grunwald Giemsa (Z)	cytoplasm, pink
	Periodic acid Schiff (B, F, Z)	cytoplasm, colourless (Z, F)
		cytoplasm, pink (B)
	Periodic acid Schiff after digestion with ptyalin (B, F, Z)	cytoplasm, colourless (B, F, Z)
	Prussian blue (Gomori's method) (F, Z)	cytoplasm, blue with dark blue inclusions (Z); cytoplasm, blue or colourless (F)
	Hale's colloidal iron (H)	cytoplasm, yellow with blue inclusions in vacuoles
	Masson's trichrome (F, Z)	cytoplasm, pale to dark red (F, Z)
	Acid phosphatase	—
	Alkaline phosphatase	—
	Lipase	—
	Peroxidase	—
Phascolosoma scolops	Neutral red	granules +
	Iron (Berlin blue)	granules +
	Acid phosphatase	granules +
	Periodic acid Schiff	cytoplasm +
	Giemsa	—
Themiste minor	Neutral red	orange
	Iron (Berlin blue)	granules +
	Periodic acid Schiff	cytoplasm +
	Giemsa	—
Golfingia nigra	Neutral red	vacuoles, pink
Golfingia ikedai	Iron (Berlin blue)	granules +
	Periodic acid Schiff	cytoplasm +
	Giemsa	—
	Acid phosphatase	—
Siphonosoma cumanese	Periodic acid Schiff	cytoplasm +

[a] A detailed breakdown on methods used and results obtained for lipids, fatty acids, sterols, compound lipids and carbohydrates may be found in Ohuye *et al.* (1961).

— Designates no data available.

B, F, H or Z designates the following fixatives: B, Bouin's fluid; F, Formalin solution; H, Helly's fluid; Z, Zenker's fluid. The staining reactions sometimes vary using the same stain but a different fixative.

Staining reaction in granulocytes	Staining reaction in large multinuclear cells	Staining reaction in ciliated urns	Researcher
—	—	semilunar cells, − ; walls +	Ohuye et al., (1961)
—	—	semilunar cells, − ; walls +	
—	—	semilunar cells, − ; walls +	
—	—	semilunar cells, + ; walls −	
—	—	semilunar cells, − ; walls +	
acidophils, vacuoles and granules, light pink	granules +	—	Blitz (1965)
basophils, cytoplasm, pink			
small granulocytes, vacuoles and granules, blue			
acidophils, eosinophilic granules	—	—	
basophils, basophilic granules	granules, pale pink	cupola, eosinophilic; lobe cell granules, basophilic; semilunar area, bluish red	Dybas (1973)
acidophils, eosinophilic granules	granules, eosinophilic or basophilic	cupola, eosinophilic; lobe cell granules, basophilic; semilunar area, basophilic	
basophils, basophilic granules			
granules, pale pink (B, F, Z)	granules, pink and colourless (B, F, Z)	pink (B, F, Z)	
granules, pale pink, rust (B, Z)			
granules, colourless (B, F)	granules colourless (B, F, Z)	lobe cells, colourless (B), pink (Z); semilunar area, pink (B, Z)	
granules, pink (Z)			
cytoplasm, pink (F, Z)	granules, pink (F, Z)	colourless (F, Z)	
cytoplasm, yellow	granules, yellow	lobe cells and semilunar area, blue	
granules, red (F, Z)	granules, red and green (F, Z)	cupola, red (F, Z); lobe cell granules, dark red (F), green (Z); ciliated cells, green (F, Z); semilunar area, green (F, Z)	
+	−	−	
+	+	cupola +	
+	−	lobe cells +	
+	+	+	
pink	cytoplasm +	—	Ochi (1975[a])
—	—	—	
cytoplasm +	—	—	
small granules, basophilic	peripheral granules, basophilic vesicle, transparent	—	
orange	—	—	
granules, acidophilic	—	—	
—	—	—	
—	—	—	
large granules, basophilic	—	—	
granules +	—	—	
—	—	—	

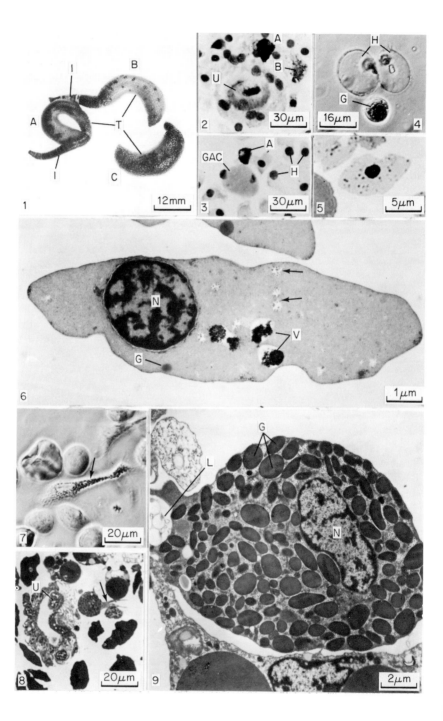

stains and localization of sub-cellular components should aid in recognition and understanding of the involvement of specific cell types in physiological and pathological functions such as haemopoiesis and wound repair.

A. Haemocytes

Haemocytes (Figs 3–7, 20) are nucleated cells containing the respiratory pigment haemerythrin dispersed throughout the cytoplasm. The cytoplasm also contains a few mitochondria and clusters of dense particles. Membrane-bound vacuoles, visible with both the light and electron microscope, contain a dark, flocculent material which gives a positive reaction for ferritin. A few membrane-bound granules are usually visible near the cell membrane (Dybas, 1973; Ochi, 1975a).

B. Granulocytes

Granulocytes (Figs 2, 3, 4, 7–10), also called amoebocytes or leukocytes, are characterized by their amoeboid nature and by numerous cytoplasmic granules. These granules vary in size, have different staining properties with Wright's blood stain and are either acidophilic or basophilic, indicating internal diversity. Other organelles present include a Golgi complex, rough

---

Fig. 1. Whole sipunculans, A and B with introverts extended, C with introvert retracted exhibiting the characteristic "peanut" shape. T, trunk; I, introvert.

Fig. 2. Whole, air-dried coelomocytes. A, acidophilic granulocyte; B, basophilic granulocyte; U, ciliated urn. Wright's stain.

Fig. 3. Whole, air-dried coelomocytes. H, haemocytes; A, acidophilic granulocyte; GAC, granulocyte with associated cell or "cell within a cell". Wright's stain.

Fig. 4. Unfixed coelomocytes. H, haemocytes; G, granulocyte. In the haemocytes, the nuclei and dense material in the cytoplasmic vacuole are visible. Phase contrast.

Fig. 5. Haemocyte with dense material in the cytoplasmic vacuoles. Thick plastic, toluidine blue.

Fig. 6. Fine structure of a typical haemocyte. N, nucleus; G, granule; V, vacuoles containing the dense material, arrows—clumped, non membrane-bound particles.

Fig. 7. Unfixed granulocyte showing the nucleus and cytoplasmic granules (arrow). The other cells in the field are haemocytes. Phase contrast.

Fig. 8. Granulocyte with pseudopodia (arrow). U, ciliated urn. Thick plastic, toludine blue.

Fig. 9. Fine structure of a typical granulocyte. N, nucleus; G, granules. Pseudopodia are surrounding latex beads (L).

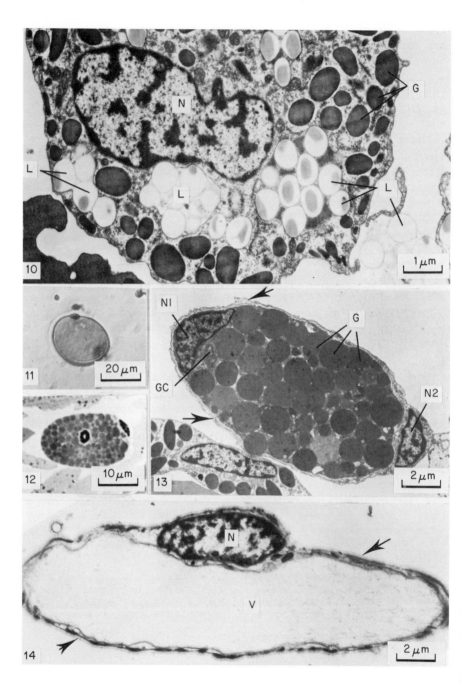

10

N

G

L

L

L

1 μm

11

20 μm

12

10 μm

13

N1

GC

G

N2

2 μm

14

N

V

2 μm

and smooth endoplasmic reticulum, ribosomes, mitochondria, lipid droplets and residual bodies. Vacuoles sometimes enclose cell organelles and an occasional haemocyte suggesting that these cells are actively phagocytic (see section on functions of coelomocytes). New granules have been seen forming from the Golgi complex (Dybas, 1973; Ochi, 1975a).

C. Large multinuclear cells

Large multinuclear cells (Figs 3, 11–14) have been reported in *P. agassizii* as being two cells, a "cell within a cell" (Dybas, 1975). Under the electron microscope, a cell containing numerous granules is seen to be completely encircled by another cell (Fig. 13). This cell pair is named "granulocyte with associated cell" or GAC. The two cells are separated by an extracellular space containing fibrils. Both cells contain a full complement of organelles: Golgi complex, mitochondria, granules and ribosomes. Granules of the inner cell have been seen forming from the Golgi complex. Under the light microscope, only the nuclei and the granules of the inner cell are visible (Fig. 12). In other species, multinuclear cells surround a hollow cavity or vesicle (Cuénot, 1900; Hérubel, 1908; Thomas, 1932–33; Volkonsky, 1933; Hyman, 1959; Ochi, 1975a). In *Golfingia gouldii* (Fig. 14), with the electron microscope, the central vesicle appears loosely filled with banded fibres. The cells surrounding the vesicle contain granules, mitochondria, and dense plaques resembling tonofilaments. The external cell membrane is also covered with villi-like projections (Fig. 14) (Dybas, unpublished).

Fig. 10. Fine structure of a granulocyte after phagocytosis of latex beads. N, nucleus; G, granules; L, latex beads.

Fig. 11. Unfixed granulocytes with associated cell. The nucleus of the external cell always appears slightly elevated from the cell. Phase contrast.

Fig. 12. Granulocyte with associated cell. The nuclei of both the inner and outer cells are visible in this plane of section. The granules of the inner cell completely fill the cytoplasm. Thick plastic, toluidine blue.

Fig. 13. Fine structure of a granulocyte with associated cell. Nl, nucleus of inner cell; G, granules of inner cell; GC, Golgi complex; N2, nucleus of outer cell, arrows—cytoplasm of outer cell. At this magnification the fibrils separating the outer and inner cells are not visible. The other cell in this field is a granulocyte.

Fig. 14. Fine structure of a multinuclear cell. In this section only one nucleus (N) is visible. The cytoplasm surrounds a large central vesicle (V) loosely filled with fibres. Villi-like projections cover the external cell membrane (arrows).

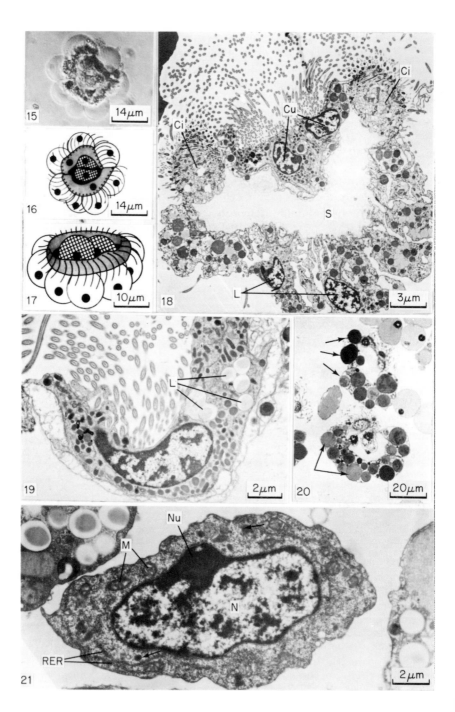

15 `14μm`

16 `14μm`

17 `10μm`

18 Ci Cu Ci S L `3μm`

19 L `2μm`

20 `20μm`

21 Nu M N RER `2μm`

D. *Ciliated urns*

Ciliated urns (Figs 2, 8, 15–20) are multicellular structures bearing cilia. In *P. agassizii*, the ciliated urns are disk shaped with cilia on one side and floral-like lobes on the opposite side (Figs 15–17). Three distinct cell types have been described: cupola cells, ciliated cells, and lobe cells (Figs 15–20). These cells are arranged around an extracellular concave disk called the semilunar area which is composed of a loose meshwork of banded fibrils. Two cupola cells containing granules are situated in the depression of the disk. Two ciliated cells are positioned on the rim of the disk surrounding the cupola cells. These cells are firmly anchored to the semilunar area by hemidesmosomes. Attached to the opposite side of the semilunar area, also by hemidesmosomes, are 10–12 lobe cells, each with its own nucleus (Figs 15–17). The cytoplasm of the lobe cells contains granules (Dybas, 1976).

The ciliated urn described for *Sipunculus nudus* has a different appearance (Bang and Bang, 1962, 1965); it is two separate cells joined by a flexible attachment. The anterior cell resembles a transparent bubble with a flattened base and short neck. The base and neck fit into the inner surface of the saucer-shaped posterior cell which has a clear central area and is ciliated on the convex outer cell surface.

Fig. 15. Unfixed ciliated urn. Phase contrast.

Fig. 16. Schematic representation of the ciliated urn viewed in Fig. 15: cupola cells, checked pattern; ciliated cells, grey; lobe cells, clear; semilunar area, black; nuclei, black dots.

Fig. 17. Schematic representation of a ciliated urn, side view (see Fig. 16 for legend).

Fig. 18. Fine structure of a ciliated urn: Cu, cupola cells; Ci, ciliated cells; L, lobe cells; S, semilunar area. In this plane of section the nuclei of the ciliated cells and some of the lobe cells are not visible.

Fig. 19. Fine structure of a ciliated urn cupola cell after phagocytosis of latex beads (L). Sections through cilia from the ciliated cells are visible above the cupola cell. A portion of the semilunar area is visible beneath the cupola cell.

Fig. 20. Two ciliated urns after incubation with ferritin particles. The lobe cells are filled with phagocytosed ferritin (arrows). Compare lobe cells with those visible in Fig. 8. The other cells visible in this field are haemocytes. Thick plastic, toluidine blue.

Fig. 21. Fine structure of an immature cell. N, nucleus; Nu, nucleolus; M, mitochondria; RER, rough endoplasmic reticulum; arrows, granules. The cytoplasm also contains numerous free ribosomes.

E. Immature cells

Immature cells (Fig. 21) which appear to be in the process of differentiating into one of the other cell types, are smaller than the other circulating cells and contain a prominent nucleolus and a cytoplasm packed with ribosomes. Occasionally small granules are visible scattered among the ribosomes (Dybas, 1973).

Data on the relative frequency of the various cell types, excluding gametes, is available for only one species, *Phascolosoma agassizii*: 90% haemocytes: 4·9% ciliated urns: 4·9% granulocytes: 0·2% large multinuclear cells—"cell within a cell" ($n = 41$) (Dybas, 1975). Immature cells were not detected.

V. Functions of coelomocytes

Sipunculans are one of four groups of animals which use the pigment haemerythrin, located in haemocytes, for transporting oxygen. Numerous studies have been performed on the function of sipunculan haemerythrin *in vitro* (e.g. Klotz, 1971; Klippenstein *et al.*, 1972; Dunn *et al.*, 1977) but only a few studies are available on the function of haemerythrin *in vivo* (Florkin, 1933; Edmonds, 1957, 1975; Mangum and Kondon, 1975).

Classification of the coelomocytes has been made by their ability to clear the coelom of foreign material by phagocytosis. *In vivo* and *in vitro* studies (Cuénot, 1900; Hérubel, 1908; Cantacuzène, 1922; Volkonsky, 1933; Ohuye, 1937; Triplett *et al.*, 1958; Towle, 1962; Blitz, 1965; Bang, 1967; Cushing *et al.*, 1970) revealed that the ciliated urns seem to agglutinate or trap the foreign particles, but the granulocytes are the primary phagocytic cells (Figs 9, 10). Dybas (1973) has demonstrated that the granules of granulocytes contain acid and alkaline phosphatase, peroxidase and lipase—enzymes that have been implicated in killing bacteria or in intracellular digestion in vertebrate leucocytes. The granules fuse with phagocytic vacuoles then empty their contents into the vacuoles. Granulocytes have also been shown to encapsulate a variety of materials inserted in the coelomic cavity which were too large to be ingested by phagocytosis (Triplett *et al.*, 1958).

Storch and Mortiz (1970) have reported that amoebocytes (see Table I) are the cells responsible for building up the musculature in the regeneration of the introvert following amputation.

The ciliated urn in *Sipunculus nudus* secretes a mucus in the presence of foreign substances (Bang and Bang, 1965). The mucus traps foreign particles and cells and the urns can often be seen dragging tails of mucus. Different types of mucus and different rates of secretion are produced in response to various substances (Bang and Bang, 1965, 1971, 1972, 1974, 1975, 1976a, 1976b, 1978; Bang, 1972; Nicosia, 1979). In *Phascolosoma agassizii*, the urn does not secrete mucus but particles and cells are trapped in the current created by the beating cilia. Both cupola cells and lobe cells are phagocytic (Figs 19, 20). The cupola cells ingest large particles—bacteria and latex beads (Dybas, 1976), whereas the lobe cells accumulate small particles—dye labelled bovine serum albumin (Blitz, 1965) and ferritin (Dybas, 1976).

The coelomic fluid has been shown to possess strong antibacterial properties (Bang and Krassner, 1958; Bang and Bang, 1962; Krassner, 1963; Rabin and Bang, 1964; Blitz, 1965; Johnson and Chapman, 1970; Krassner and Florey, 1970). Injection of bacteria results in production of a non-specific bactericidin that is not present, or only present in small amounts, in the non-injected animals (Evans et al., 1969, 1973). Injection of either ciliates, bacteria or erythrocytes causes rapid production of lysins for these substances (Cantacuzène, 1922; Blitz, 1965; Bang, 1966, 1967; Cushing et al., 1969; Weinheimer et al., 1970). A substance called stop factor, present in the coelomic fluid, can immobilize a marine dinoflagellate within ten minutes; recovery takes twelve hours (Cushing et al., 1969). The bactericidins, lysins and agglutinins can be separated from one another on the basis of characteristic properties such as thermostability, enzyme sensitivity and filtration. It is not known which, if any, coelomocytes are involved in the production of these substances.

VI. Summary and concluding remarks

Five types of coelomocytes have been reported for sipunculans: (1) haemocytes which carry the respiratory pigment haemerythrin and function in oxygen transport; (2) granulocytes, the main phagocytic cells; (3) large multinuclear cells whose function is unknown; (4) ciliated urns, which trap foreign particles in mucus secretions in one species, phagocytose and trap particles in another species and aid in movement of coelomic fluid; and (5) immature cells, which appear to be stem cells capable of differentiating into other cell types. No haemopoietic organs have been identified.

More data is needed on the structure and function of the large multinuclear cells and ciliated urns. These cell types appear to vary considerably from species to species. The possible involvement of coelomocytes in wound healing and clotting, and bactericidin, agglutinin and lysin production needs to be investigated as well as the origin and life span of these unusual cells.

PART 2: ECHIUROIDS

I. Introduction

Echiura, commonly called spoonworms, are a small phylum of marine worms consisting of 130 described species (Stephen and Edmonds, 1972). Echiuroids are similar in appearance to sipunculans but differ by having an extendable, but non-retractable, grooved proboscis with a mouth at its base (Fig. 22). Most echiuroids live in burrows or crevices in rocks; some build U-shaped tubes. They are usually sedentary animals but can move around when necessary. They use their proboscis for exploration of their surroundings and for feeding. Food particles stick to mucus secretions from the proboscis, then muscular and ciliary action moves the mucus masses along the groove in the proboscis to the mouth (Jaccarini and Schembri, 1977). A muscular body encloses an undivided coelomic cavity containing one or more nephridia, a convoluted digestive tube, and a nerve cord, but no definite brain or ventral ganglia (Sims, 1976; Meglitsch, 1967).

Sexes are separate, but in one species the male lives inside the female and fertilizes the eggs before they are laid (MacGinitie and MacGinitie, 1949). Gametes arise from gonads along the ventral mesentery as they do in sipunculans. Gametes are shed into the coelom where they mature, are sequestered in a storage organ, and are finally released through the nephridia. Fertilized eggs undergo spiral cleavage and become trochophore larvae. The embryos pass through a transient segmented stage (Meglitsch, 1967; Sims, 1976) which is the basis for placing echiuroids in a separate phylum from annelids, the segmented worms, and sipunculans which exhibit no segmentation in either the larval or adult stages.

II. Structure of the circulatory system

The coelomic cavity in the body of echiuroids is continuous. In the proboscis, modifications and divisions of the coelom vary with species (Menon, 1975). Most species also have a closed vascular system comprising dorsal, ventral and neurointestinal blood vessels. The dorsal vessel arises at the posterior circumintestinal vessel and runs to the tip of the proboscis where it branches. The branches re-join behind the mouth to form the ventral vessel which runs back above the nerve cord (Meglitsch, 1967; Menon and Dattagupta, 1975). In some species, including the most studied echiuroid, *Urechis caupo*, the circulatory system is absent. Some respiration takes place across the body wall but the thin hind gut is the main site of oxygen exchange. Water is drawn through the anus into the hind gut by peristaltic movements of the body wall. The water is expelled intermittently and replaced with fresh sea water (MacGinitie and MacGinitie, 1949). Simple diffusion of oxygen is adequate for most metabolic processes. The coelomocytes with haemoglobin transport oxygen primarily during periods of low tide and reduced oxygen availability (Redfield and Florkin, 1931). In some species, crystalline haemoglobin, enough to supply the animal with oxygen for one hour, has been found in the plasma (MacGinitie and MacGinitie, 1949). Coelomocytes are present in both the coelomic cavity and the blood vessels. In one species, however, chloragogen cells are found in the blood vessels and only rarely in the coelomic cavity (Menon, 1975).

III. Origin and formation of coelomocytes

The origin of coelomocytes has been reported in only three species of echiuroids (Menon, 1975). In *Rubriccelatus pirotansis* the mesentery in the presiphonal intestine contains active centres of proliferation of coelomocytes. Chloragogen cells have been reported to proliferate from the peritoneal lining of the alimentary canal and blood vessels in *Listriolobus brevirostris* and *Ochetostoma septemyotum*.

IV. Structure and classification of coelomocytes

A. Erythrocytes

Erythrocytes (Figs 23–26) are the most numerous cell type and contain the respiratory pigment haemoglobin which is similar, but not identical, to vertebrate haemoglobin (Redfield and Florkin, 1931). After sedimentation, the volume of packed erythrocytes is 18–40% of the total volume of coelomic fluid. Other cells, accounting for about 1% of the total packed cell volume, form a thin buffy coat between the erythrocytes and the remaining plasma (Redfield and Florkin, 1931; MacGinitie, 1935). The erythrocytes are unusual because they differ morphologically in males, females, and in sexually immature animals in one species (Baumberger and Michaelis, 1931) and with different stages of maturation in another species (Ochi, 1965) (Table III). Brown or black pigment and crystals are found in the cytoplasm of living erythrocytes (Table III; Fig. 23). The cytoplasm varies in staining properties and in the size and number of granules (Tables III and IV). With histochemical stains the granules have been determined to be lysosomes, iron deposits, and lipid granules (Ohuye, 1937, 1938a,b; Ochi, 1966, 1975b). Glycogen is scattered throughout the cytoplasm (Ochi,

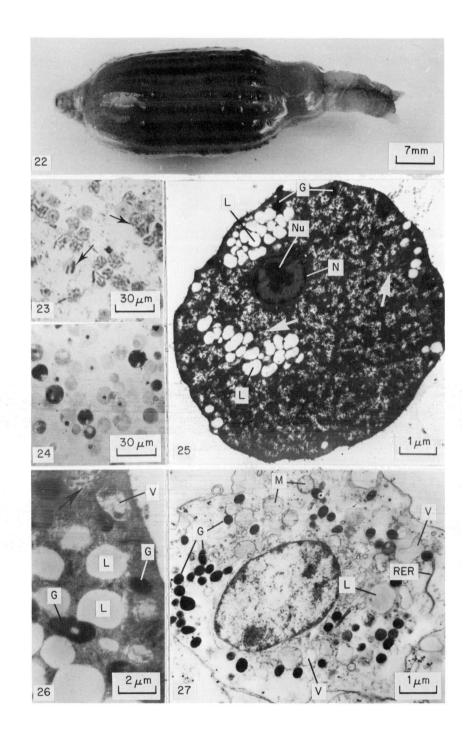

1966, 1975b). Representative cytoplasmic organelles can be seen in Figs 25 and 26. The ease with which certain cytoplasmic components are extracted during processing of coelomocytes for histochemical studies and electron microscopy makes correlation of whole cells with their fixed, embedded, sectioned and stained counterparts difficult.

B. Amoebocytes

Amoebocytes (Fig. 27, Table III) contain yellow or brown pigment in the living state and are amoeboid. Varying numbers of granules are scattered throughout the cytoplasm. Detailed drawings of these cells may be found in Abbott (1913).

C. Chloragogen cells

Chloragogen cells which are not illustrated, are flask-shaped cells containing one or more inclusions of different size and shape. These cells are usually associated with the blood vessels, siphon and specialized portions of the peritoneum (Menon, 1975).

Fig. 22. A typical echiuroid (*Thalassema* sp., Kahalui, Hi. U.S.A.) showing the characteristic plump body and grooved proboscis.

Fig. 23. A representative sample of air-dried, whole coelomocytes of *Urechis caupo*. The erythrocytes vary in size and shape. Note haemoglobin crystals (arrows). Wright's stain.

Fig. 24. Erythrocytes of *U.* caupo. The cells show variation in staining and in number of lipid granules in the cytoplasm. Haemoglobin crystals, so prominent in whole, air-dried coelomocytes, were never observed in cells after fixation and dehydration in graded alcohols. Thick plastic, toluidine blue.

Fig. 25. Fine structure of a typical erythrocyte from *U.* caupo. A nucleus (N) with a prominent nucleolus (Nu) and cytoplasm containing lipid droplets (L), electron-dense granules (G), and abundant glycogen (arrows) are visible. The lipid has been extracted during processing for electron microscopy.

Fig. 26. The characteristic cytoplasmic components of an erythrocyte of *U. caupo* at higher magnification: lipid droplets (L), electron-dense granules (G), membrane-bound vacuole (V) and glycogen (arrow).

Fig. 27. Fine structure of an amoebocyte of *U. caupo*. The cytoplasm contains numerous electron-dense granules (G), large mitochondria (M), membrane-bound vacuoles (V), lipid droplets (L), and rough endoplasmic reticulum (RER). The cytoplasmic ground substance is poorly preserved.

Table III. Classification of Echiura coelomocytes.

Genus, species	Erythrocytes	Amoebocytes and other cells	Researcher
Echiurus chilensis *Echiurus echiurus*	erythrocytes with haemoglobin[a]	cells with yellow pigment granules[a]	Seitz (1907)
Thallassema mellita	hematids with brown pigment, intracellular haemoglobin crystals, few to many granules (8–9 μm)	1. cells stick to glass (3–5 μm) 2. cells with blackberry appearance, pseudopodia (12–14 μm) 3. multinucleate cells with granules and yellowish inclusions[a]	Abbott (1913)
Urechis caupo	erythrocytes (25 μm)	yellow amoeboid cells[a]	Fisher and MacGintie (1928)
Urechis caupo	erythrocytes[a] in immature animals—brown granules; in mature males—black granules, scattered brown pigment, no brown granules; in mature females—no black granules, disintegrating brown granules;	yellow mulberry droplets	Baumberger and Michaelis (1931)
Thallassema gogoshimense	erythrocytes with haemoglobin crystals (10–36 μm × 3–6 μm)	1. hyaline amoebocytes[a] 2. finely granular amoebocytes[a] 3. coarsely granular amoebocytes[a] 4. compartmental amoebocytes[a]	Ohuye (1937, 1938a,b)
Urechis unicicutus	(25 μm × 4 μm)		

Review	erythrocytes with haemoglobin[a]	amoebocytes with pseudopodia and brown pigment[a]	Grassé (1959)
Urechis unicicutus	erythrocytes[a] in immature animals—lipid, mitochondria; in small, mature animals— lipid, mitochondria, dense bodies; in large, mature animals—lipid, mitochondria, dense bodies, pigment granules;	—	Ochi (1965, 1966, 1975b)
Ochetostoma capensis *Ochetostoma septemyotum* *Ochetostoma zanzibarensis* *Listriolobus brevirostris* *Anelassorhynchus branchio-rhynchus*	spherical cells[a]	1. rayed cells[a] 2. amoeboid cells[a]	Menon (1975)
Thalassema neptuni *Rubricelatus pirotansis*	spherical cells[a]	amoeboid cells[a] 1. multinucleate corpuscles[a] 2. chloragogen cells with inclusions[a]	

[a] Size not measured; — no data available.

The literature relating to classification of echiuroid coelomocytes is summarized in Table III. The literature pertaining to the staining properties of the coelomocytes is reviewed in Table IV.

V. Functions of coelomocytes

Echiuroid erythrocytes contain larger amounts of lipid and glycogen than are found in the erythrocytes of most vertebrates and invertebrates. Since their oxygen carrying capacity is used only as a back-up system (Ditadi, 1975), their main function is probably transportation of other nutrients (Ochi, 1966; Lawrence *et al.*, 1971; Marsden, 1976). The lysosomes found in the cytoplasm of erythrocytes may be involved in the breakdown of lipid granules (Ochi, 1965). Eggs injected into the coelomic cavity of males are fertilized and develop more rapidly than those raised in culture dishes (MacGinitie, 1935) confirming the abundance of nutrients in the coelomic fluid. Studies relating to the function of the amoebocytes and chlorogogen cells are lacking.

In a preliminary study using the blood of *Urechis caupo*, coelomocytes clumped in the presence of bacteria but no phagocytosis of bacteria by coelomocytes could be observed with either the light or electron microscope. Samples of coelomic fluid were incubated *in vitro* with three different species of bacteria (*Escherichia coli*, *Staphylococcus aureus* and *Proteus vulgaris*). The coelomic fluid, either with or without coelomocytes, exhibited a marked bactericidal effect over the first four hours incubation (Dybas, unpublished data).

VI. Summary and concluding remarks

The coelomic fluid of echiuroids contains three coelomocyte types: erythrocytes, amoebocytes and chloragogen cells. Erythrocytes, the most numerous cell type, contain haemoglobin and large amounts of lipid and glycogen. These cells transport lipids, carbohydrates and oxygen during periods of limited oxygen availability. In some species, the cytoplasmic constituents vary with sex and age of the animals. Amoebocytes and chlorogogen cells are present in low numbers; their function is unknown. No data is available regarding function or involvement of coelomocytes in healing and repair or in defence. These areas should prove profitable avenues for future research.

TABLE IV. Staining properties of Echiura coelomocytes.

Genus, species	Stain	Staining reaction in erythrocytes[a]	Researcher
Thallassema gogoshimense			Ohuye (1937,
Urechis unicicutus	Nile blue sulphate	+	1938a, 1938b)
	Neutral red	+	
	Oxidase	+	
	Sudan III	+	
	Iron	+	
	Silver impregnation	+	
	Osmic acid	+ (mitochondria)	
	Janus green	basophilic cytoplasm	
	Giemsa	+	
	Vitamin A	+	
	Vitamin C		
Urechis unicicutus	Sudan III	+ (lipid granules)	Ochi (1965, 1966)
	Sudan black	+ (lipid granules)	
	Berlin blue (iron)	+ (dense bodies)	
	Acid phosphatase	+ (colourless granules)	
	Periodic acid Schiff (glycogen)	+ (cytoplasm)	
	Periodic acid Schiff (digestase)	− (cytoplasm)	

[a] No data is available for amoebocytes.

References

Abbott, J. F. (1913). *Wash. Univ. Studies* **1**, 3–9.
Åkesson, B. (1958). *Undersökningar över Öresund* **38**, 1–249.
Andrews, E. A. (1890). *Stud. Biol. Lab. Johns Hopkins Univ.* **4**, 389–430.
Bang, B. G. (1972). *Fed. Proc.* **31**, 663.
Bang, B. G. and Bang, F. B. (1965). *Cah. Biol. Mar.* **6**, 257–264.
Bang, B. G. and Bang, F. B. (1971). *Cah. Biol. Mar.* **12**, 3–10.
Bang, B. G. and Bang, F. B. (1972). *Am. J. Pathol.* **68**, 407–422.
Bang, B. G. and Bang, F. B. (1974). *The Lancet* **2**, 1292–1294.
Bang, B. G. and Bang, F. B. (1975). *Nature (Lond.)* **253**, 634–635.
Bang, B. G. and Bang, F. B. (1976a). *Cah. Biol. Mar.* **17**, 423–432.
Bang, B. G. and Bang, F. B. (1976b). *Fed. Proc.* **35**, 763.
Bang, B. G. and Bang, F. B. (1978). *Fed. Proc.* **37**, 266.
Bang, F. B. (1966). *J. Immunol.* **96**, 960–972.
Bang, F. B. (1967). *Fed. Proc.* **26**, 1680–1684.
Bang, F. B. and Bang, B. G. (1962). *Cah. Biol. Mar.* **3**, 363–374.
Bang, F. B. and Krassner, S. B. (1958). *Biol. Bull., Woods Hole* **158**, 343.
Baumberger, J. P. and Michaelis, L. (1931). *Biol. Bull., Woods Hole* **61**, 417–421.
Blitz, R. R. (1965). Ph.D. Thesis, Univ. California, Berkeley, U.S.A.
Brown, A. C. and Winterbottom, R. (1969). *J. Invertebr. Pathol.* **13**, 229–234.
Cantacuzène, J. (1922). *C.R. Soc. Biol. (Paris)* **87**, 259–263.
Cuénot, L. (1900). *In* "Zoologie Descriptive des Invertebres" (L. Boutan, Ed.), Vol. 1, pp. 386–422. O. Doin, Paris.
Cushing, J. E., McNeely, J. L. and Tripp, M. R. (1969). *J. Invertebr. Pathol.* **14**, 4–12.
Cushing, J. E., Tripp, M. R. and Fuzessery, S. (1970). *Fed. Proc.* **29**, 771.
Dawson, A. B. (1933). *Biol. Bull., Woods Hole* **64**, 233–242.
Ditadi, A. S. F. (1975). *In* "Proc. Int. Symp. Biol. Sipuncula and Echiura" (M. E. Rice and M. Todorović, Eds), Vol. II, pp. 191–196. Naučno Delo, Beograd.
Dunn, J. B. R., Addison, A. W., Bruce, R. E., Loehr, J. S. and Loehr, T. M. (1977). *Biochemistry* **16**, 1743–1749.
Dybas, L. (1973). M.A. Thesis, California State Univ., San Francisco, U.S.A.
Dybas, L. (1975). *Nature (Lond.)* **257**, 790–791.
Dybas, L. (1976). *Cell Tiss. Res.* **169**, 67–75.
Edmonds, S. J. (1975). *In* "Proc. Int. Symp. Biol. Sipuncula and Echiura" (M. E. Rice and M. Todorović, Eds), Vol. I, pp. 3–9. Naučno Delo, Beograd.
Edmonds, S. J. (1957). *Aust. J. Mar. Freshwater Res.* **8**, 55–63.
Evans, E. E., Weinheimer, P. F., Acton, R. T. and Cushing, J. E. (1969). *Nature (Lond.)* **222**, 695.
Evans, E. E., Cushing, J. E. and Evans, M. L. (1973). *Infect. Immunity* **8**, 355–359.
Fisher, W. F. and MacGinite, G. E. (1928). *Ann. Mag. Nat. Hist.* **1**, 199–204.
Florkin, M. (1933). *Arch. Int. Physiol.* **36**, 247–328.
Grassé, P. P. (1959). *In* "Traité de Zoologie", Vol. 5, pp. 854–907. Masson et Cie, Paris.
Hérubel, M. A. (1908). *Mem. Soc. Zool. (Paris)* **20**, 107–418.
Hyman, L. H. (1959). *In* "The Invertebrates", Vol. 5, pp. 610–696. McGraw Hill, New York.
Jaccarini, V. and Schembri, P. J. (1977). *J. exp. mar. Biol. Ecol.* **28**, 163–181.
Johnson, P. T. and Chapman, F. A. (1970). *J. Invertebr. Pathol.* **16**, 127–138.

Klippenstein, G. L., Van Riper, D. A. and Oosterom, E. A. (1972). *J. biol. Chem.* **247**, 5959–5963.

Klotz, I. M. (1971). *In* "Subunits in Biological Systems" (S. N. Timasheff and G. D. Fasman, Eds), pp. 55–103. Marcel Dekker, New York.

Krassner, S. M. (1963). *Biol. Bull., Woods Hole* **125**, 382–383.

Krassner, S. M. and Florey, B. (1970). *J. Invertebr. Pathol.* **16**, 331–338.

Lankester, E. W. (1908). *Nature (Lond.)* **78**, 318.

Lawrence, A. F., Lawrence, J. M. and Giese, A. C. (1971). *Comp. Biochem. Physiol.* **38B**, 463–465.

MacGinitie, G. E. (1935). *J. exp. Zool.* **71**, 483–487.

MacGinitie, G. E. and MacGinitie, N. (1949). *In* "Natural History of Marine Animals", pp. 184–193. McGraw Hill, London.

Mangum, C. P. and Kondon, M. (1975). *Comp. Biochem. Physiol.* **50**, 777–785.

Marsden, J. R. (1976). *Comp. Biochem. Physiol.* **53B**, 225–229.

Meglitsch, P. (1967). *In* "Invertebrate Zoology", pp. 645–648. Oxford University Press, London.

Menon, P. K. B. (1975). *In* "Proc. Int. Symp. Biol. Sipuncula and Echiura" (M. E. Rice and M. Todorović, Eds), Vol. II, pp. 183–190. Naučno Delo, Beograd.

Menon, P. K. B. and Dattagupta, A. K. (1975). *In* "Proc. Int. Symp. Biol. Sipuncula and Echiura" (M. E. Rice and M. Todorović, Eds), Vol. II, pp. 177–182. Naučno Delo, Beograd.

Nicosia, S. V. (1979). *Science* **206**, 698–700.

Ochi, O. (1965). *Mem. Ehime Univ. Sect. II, Ser. B* **V**, 35–42.

Ochi, O. (1966). *Mem. Ehime Univ. Sect. II, Ser. B* **V**, 71–75.

Ochi, O. (1975a). *In* "Proc. Int. Symp. Biol. Sipuncula and Echiura" (M. E. Rice and M. Todorović, Eds), Vol. I, pp. 219–227. Naučno Delo, Beograd.

Ochi, O. (1975b). *In* "Proc. Int. Symp. Biol. Sipuncula and Echiura" (M. E. Rice and M. Todorović, Eds), Vol. II, pp. 197–204. Naučno Delo, Beograd.

Ohuye, T. (1937). *Sc. Rep. Tohoku Imp. Univ.* **12**, 203–239.

Ohuye, T. (1938a). *Sc. Rep. Tohoku Imp. Univ.* **12**, 623–628.

Ohuye, T. (1938b). *Sc. Rep. Tohoku Univ.* **13**, 359–380.

Ohuye, T., Ochi, O. and Miyata, I. (1961). *Mem. Ehime Univ. Sect. II, Ser B* **5**, 145–151.

Rabin, H. and Bang, F. B. (1964). *J. Insect Pathol.* **6**, 457–465.

Rajulu, G. S. and Krishnan, N. (1969). *Nature (Lond.)* **223**, 186–187.

Redfield, A. C. and Florkin, M. (1931). *Biol. Bull., Woods Hole* **61**, 185–210.

Rice, M. E. (1970). *Science* **167**, 1618–1620.

Rice, M. E. (1973). *Smith. Contrib. Zool,* **132**, 1–51.

Seitz, P. (1907). *Zool. Jahrb. Anat.* **24**, 323–356.

Selensky, W. (1908). *Zool. Anz.* **32**, 329–336.

Sims, R. W. (1976). *In* "Encyclopedia Britannica", Vol. 6, pp. 186–187. Encyclopaedia Britannica, Inc., Chicago.

Stang-Voss, C. (1970). *Z. Zellforsch. mikrosk. Anat.* **106**, 200–208.

Stephen, A. C. and Edmonds, S. J. (1972). "The Phyla Sipuncula and Echiura", The British Museum (Natural History), London.

Storch, V. and Moritz, K. (1970). *Z. Zellforsch. mikrosk. Anat.* **110**, 258–267.

Thomas, J. A. (1931). *C.R. Acad. Sciences (Paris)* **193**, 1462–1465.

Thomas, J. A. (1932a). *C.R. Soc. Biol. (Paris)* **110**, 451–452.

Thomas, J. A. (1932b). *Ann. Inst. Pasteur* **49**, 234–263.

Thomas, J. A. (1932–33). *Arch. Zool. exp. gén.* **13**, 22–40.

Towle, A. (1962). Ph.D. Thesis. Stanford Univ., California, U.S.A.

Towle, A. (1975). *In* "Proc. Int. Symp. Biol. Sipuncula and Echiura" (M. E. Rice and M. Todorović, Eds), Vol. I, pp. 211–218. Naučno Delo, Beograd.

Triplett, E. F., Cushing, J. E. and Durall, G. L. (1958). *Amer. Naturalist* **92,** 287–293.

Volkonsky, M. (1933). *Biol. Bull. (France et Belgium)* **67,** 135–199.

Weinheimer, P. F., Acton, P. T., Cushing, J. E. and Evans, E. E. (1970). *Life Sci.* **9,** 145–152.

Yeager, J. F., Franklin, J. and Tauber, O. E. (1935). *Biol. Bull., Woods Hole* **69,** 66–70.

Section IV

Molluscs

7. Gastropods

T. SMINIA

Vrije Universiteit, Biologisch Laboratorium, Amsterdam—1007mc, The Netherlands

CONTENTS

I. Introduction

The morphology and functions of the cells present in the haemolymph of gastropod molluscs have been studied by many workers (e.g. Tripp, 1970; Sminia, 1972; Sminia et al., 1973, 1974; Yoshino, 1976; Cheng and Garrabrant, 1977). Although species from all three gastropod subclasses (Prosobranchia, Opisthobranchia and Pulmonata) have been investigated, most data on blood cell structure and function have been obtained from studies of pulmonates, in particular, the terrestrial snails, *Helix aspersa* and *H. pomatia* (e.g. Wagge, 1955; Prowse and Tait, 1969; Bayne, 1974) and the fresh water snails, *Biomphalaria glabrata* (e.g. Pan, 1958; Tripp, 1961; Faulk et al., 1973; Cheng and Auld, 1977) and *Lymnaea stagnalis* (e.g. Müller, 1956; Stang-Voss, 1970; Sminia, 1972, 1977a). As yet, no studies have been made on blood cells of terrestrial slugs.

The blood cells of gastropods are often called leucocytes; other names used are haemocytes, amoebocytes, granulocytes, lymphocytes and macrophages (e.g. Wagge, 1955; Kress, 1968; Cheng et al., 1969; Davies and Partridge, 1972; Cheng and Auld, 1977). None of the species investigated possesses blood cells which are morphologically and functionally comparable to vertebrate erythrocytes. The respiratory (blood) pigments—haemoglobin in planorbid snails and haemocyanin in the other gastropods—are present in the serum. It has been shown that these blood pigments are synthesized, stored and released by a special type of connective tissue cell, the pore cell (Sminia, 1972; Sminia et al., 1972; Sminia and Boer, 1973; Sminia, 1977b; Sminia and Vlugt-van Daalen, 1977).

It has become evident that in gastropods, blood cells play a major part in the internal defence system (e.g. Tripp, 1970; Harris and Cheng, 1975a,b; Sminia, 1977a). For this reason, parasitologists have paid much attention to the blood cells of snails, which are intermediate hosts of parasites injurious to mammals including man (e.g. Kinoti, 1971; Meuleman, 1972; Carter and Bogitsh, 1975; Harris, 1975). Other functions attributed to gastropod blood cells are involvement in tissue repair (e.g. Armstrong et al., 1971; Sminia et al., 1973), digestion, transport of various substances and shell repair (e.g. Wagge, 1955; Aboliņš-Krogis, 1973, 1976). Apparently, gastropod blood cells are multipotent cells.

Following a general description of the morphology of the circulatory system of gastropods, the rest of the chapter will deal with structure and function of blood cells, and in particular blood cell functions in phagocytosis, encapsulation, wound healing and haemocytopoiesis.

II. Structure of the circulatory system

Gastropods have an open circulatory system. The heart pumps the haemolymph into the arterial system which continues into arterioles and capillaries running upon or inside the organs. The haemolymph is then collected in venous slits and in sinuses in the head/foot and visceral region. From these sinuses, which are in open connection with the connective tissue, the haemolymph is collected in large venous vessels, whence it flows to the kidney region and to the mantle skirt and the gill(s) or lung and then returns to the heart. The general structure of the heart is dependent on that of the pallial organs. In gastropods with a single ctenidium or with a lung, the heart has two chambers, a ventricle and an auricle; gastropods with paired ctenidia have two auricles (e.g. Fretter and Graham, 1965; Hill and Welsh, 1966).

The gross-morphology of the circulatory system can be readily studied by injecting specimens with dyes such as India ink, or, for permanent preparations, with materials such as latex or dentist's cement. Studies of this type have mainly been performed on pulmonate snails (*H. pomatia*, Schmidt, 1916; Nold, 1924; *Agriolimax reticulatus*, Laryea, 1970; *B. glabrata*, Pan, 1971; *L. stagnalis*, Bekius, 1972) and for this reason the circulatory system of a pulmonate is presented as a model. Two diagrams show the main arteries, veins and sinuses of *L. stagnalis* (Figs 1, 2; see Bekius, 1972).

A. The arterial system

Figure 1 shows that the heart of *L. stagnalis*, consisting of an auricle and a ventricle, is located on the left side of the body, in front of the kidney. The auricle receives haemolymph from the vena renopulmonalis and the ventricle pumps it into the aorta. Auricle and ventricle are separated by a tubular valve, which projects into the ventricle (Plesch *et al.*, 1975). A crescent-shaped aortic valve prevents back-flow of the haemolymph to the ventricle. In the arteries local widenings occur. These "arterial" sinuses are centres of ramification; numerous small capillaries arise from these sinuses. In some places the arterial wall is relatively thick. Probably these parts of the arteries assist in pumping the haemolymph to the organs.

B. The venous system

Figure 2 shows that haemolymph is collected in two large haemocoelic

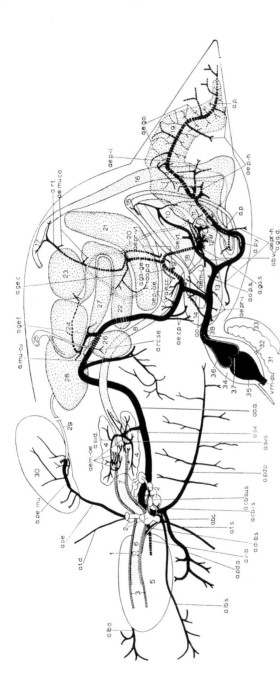

Fig. 1. The main *arterial* vessels of the snail *Lymnaea stagnalis*, viewed from the left. The pericardium and the kidney are deflected to the left. The digestive gland is omitted. 1, buccal ganglia; 2, ring of central ganglia; 3, salivary ducts; 4, salivary glands; 5, buccal mass; 6, pro-oesophagus; 7, mid-oesophagus; 8, post-oesophagus; 9, crop; 10, gizzard; 11, pylorus; 12, vestibulum with ducts of the digestive gland; 13, pro-intestine; 14, pellet compressor; 15, mid-intestine; 16, post-intestine; 17, rectum; 18, caecum; 19, ovotestis; 20, sperm-oviduct; 21, glandula albuminifera; 22, pars contorta oviducti; 23, muciparous gland (glandula nidamentaria accessoria); 24, oöthecal gland (pars nidamentaria oviducti); 25, pars vaginalis; 26, receptaculum seminis; 27, sperm duct; 28, glandula prostata; 29, vas deferens; 30, penial complex; 31, kidney; 32, reno-pericardial duct; 33, nephridial gland; 34, pericardium; 35, atrium; 36, ventricle; 37, atrio-ventricular valves; 38, aortic valve.

ab.c., arborescence cephalica; ab.v., arborescence vestibularis; a.cb. (i.s., su.s.), arteria cerebri (inferior sinistra, superior sinistra); a.d.-b.s., arteria dorso-buccalis sinistra; ae.cc., arteriae caecales; ae.cp., arteriae compressoriae; ae.go., arteriae gonadales; ae.m. -i., arteriae medio-intestinales; ae.m. -oe., arteriae medio-oesophageales; ae. mu.co., arteriae musculi columellaris; ae. p. -h., arteriae post-hepaticae; ae. p. -i., arteriae post-intestinales; ae.p. -oe., arteriae post-oesophageales; ae. pr. -h., arteriae pro-hepaticae; ae;pr. -i., arteriae pro-intestinalis; ae. pr. -vt., arteriae pro-ventriculares; ae-vt., arteriae ventriculares; a.ga. (d., s.), arteria gastrica (dextra, sinistra); a.ga. -ge., arteria gastro-genitalis; a.ge. (c., f.), arteria genitalis (caudalis, frontalis); a.lb. (d., s.), arteria labialis (dextra, sinistra); a. mu. -cu. arteria musculo-cutaneus; ao., aorta (a., p.d. p.s.), aorta (anterior, posterior dextra, posterior sinistra); a.p., arteria posterior; a.pd. (a., p.), arteria pedalis (anterior, posterior); a.pe., arteria penis; a.pe.mu., arteria penis muscularis; a.py., arteria pylorica; a.re.se., arteria receptaculum seminis; a.rt., arteria rectalis; a.sv. (d., s.), arteria salivaris (dextra, sinistra); a.t. (d., s.), arteria tentacularis (dextra, sinistra); v.rn. -pu., vena reno-pulmonalis [Courtesy of Dr R. Bekius (1972). *Neth. J. Zool.* **22**, 1–58].

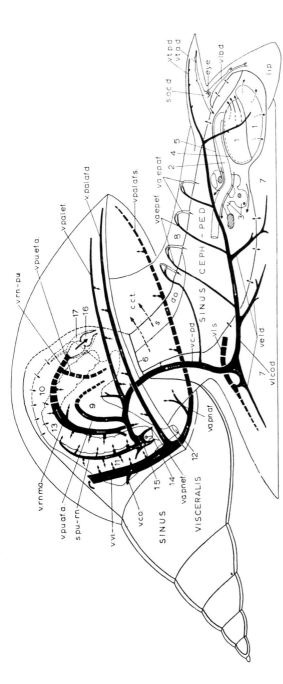

Fig. 2. The main *venous* vessels and sinuses of *L. stagnalis*, viewed from the right side. The organs and vessels shown in broken lines are situated in the left half of the body; the normal direction of the blood flow is indicated by solid arrows; the circulation at the moment of the blood extrusion is marked with broken arrows. 1, buccal mass; 2, oesophagus; 3, CNS with surrounding connective tissue; 4, membrane on upper half of the buccal mass; 5, membrana capito-cerebralis; 6, membrana transversa; 7 and 8, connective tissue of the foot and of the dorsal part of the head-foot, respectively; 9, 10 and 11, anterior, left posterior and right posterior part of the lung roof, respectively; 12, pneumostome; 13, kidney; 14, renal pore; 15, haemal pore; 16, heart; 17, pericard; c.c.t., central connective tissue; s.do., sinus dorsalis; s.oc.d., sinus oculis dextra; s.pu.-rn., sinus pulmorenalis; va.ep. (af.,ef.), vasa epidermalis (afferentia, efferentia); va.pn. (af., ef.), vasa pneumostomales (afferentia, efferentia); va.pu. (af.a., ef.a) vasa pulmonalis (afferentia anterior, efferentia anterior); v.co., vena columellaris; v.c. -pd., vena cephalo-pedalis; v.lb.d., vena labialis dextra; ve.l.d., venae laterales dextra; v.l. (co.d., s.), vena lateralis (communis dextra, sinistra); v. pal. (af.d., af.s., ef.), vena pallialis (afferens dextra, afferens sinistra, efferens) v. rn.ma., vena renalis major; v.rn. -pu., vena reno-pulmonalis; v.t. (a.d., p.d.), vena tentacularis (anterior dextra, posterior dextra); v.vi., vena visceralis. [Courtesy of Dr R. Bekius (1972). *Neth. J. Zool.* **22**, 1–58].

spaces, the sinus cephalo-pedalis and the sinus visceralis, and in the central connective tissue located around the lung cavity. The two sinuses are completely separated by a thin septum, the membrana transversa (see Fig. 2). Most of the organs of the visceral mass lie in the sinus visceralis, whilst the kidney and the accessory sex glands are mainly located in the central connective tissue. The majority of the remaining internal organs (e.g. buccal mass, oesophagus, central nervous system) are located in the sinus cephalo-pedalis. From the sinuses and the central connective tissue the haemolymph runs through a complicated system of venous vessels which often function as afferent vessels for many organs and tissues. Ultimately the short vena reno-pulmonalis collects the haemolymph and brings it back to the heart.

C. Histology of the heart and blood vessels

The heart of gastropods lies within the pericardial cavity. It is covered externally by the epicardium, a single layer of cells which is continuous with the pericardium, the cell layer lining the pericardial cavity. Both auricle and ventricle possess spongy walls, consisting of bundles of cross-striated muscle fibres (e.g. Plesch et al., 1975). There is no proper endothelium separating the muscle cells and the lumen of the heart: only scattered endothelial cells occur. It seems justified to assume that the arteries and veins also lack a continuous endothelium. Whether or not these scattered endothelial cells, which morphologically resemble fibroblasts are phagocytic is still a matter of controversy (e.g. Bayne, 1973; Sminia et al., 1979b). The walls of the vessels also contain connective tissue elements and muscle cells. The thickness and the composition of the walls vary greatly and are largely dependent on the location and the function of the vessels. For example, the walls of the aorta and of arteries are relatively thick; several veins also possess a fairly sturdy wall, comparable to that of arteries, while others totally lack a proper wall and are just spaces within connective tissue (venous slits). The connective tissue of the vessel walls consists of fibroblasts and collagen fibres, embedded in ground substance. Elastic fibres are absent (e.g. Sminia, 1972). The muscle tissue consists of smooth and/or of obliquely striated muscle cells, which occur singly or arranged in bundles (e.g. Pan, 1958; Plesch et al., 1975). In large arteries, usually an inner circular and an outer longitudinal muscle layer can be distinguished. Around the large blood vessels numerous vesicular connective tissue cells occur (Fig. 3). These glycogen storing cells (e.g. Sminia, 1972; Meuleman, 1972) are turgid and thus resistant to deformation; due to this property they give firmness to the wall of these vessels (e.g. Baecker, 1932; Carriker and Bilstad, 1946).

D. Collection of haemolymph

Haemolymph can be obtained by puncturing the heart or the large blood vessels. In several shell-bearing terrestrial and freshwater pulmonates (e.g. *Zebrina dedrita*, Gebhart-Dunkel, 1953; *Planorbis corneus*, Wesenburg-Lund, 1939; *L. stagnalis*, Müller, 1956) another, quite easy, way to obtain haemolymph is to touch the snail's foot. In doing so, the snail retracts very deeply into its shell and extrudes several drops of fluid. It has been concluded for *L. stagnalis* that this fluid, which is extruded via the haemal pore, an opening in the wall of the mantle cavity near the anus (Lever and Bekius, 1965), can be regarded as haemolymph. This conclusion is based on the observations that (1) the blood cell density of this fluid equals that of the haemolymph taken directly from the vena pulmonalis or from the heart of the same specimen (Sminia, 1972), and (2) this fluid and haemolymph obtained by direct bleeding have the same ionic composition (de With, personal communication).

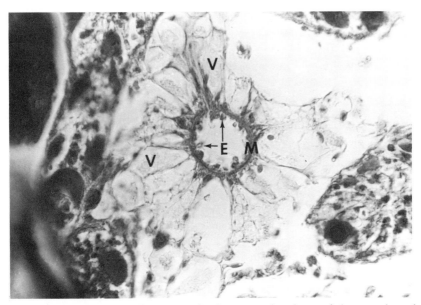

Fig. 3. Light micrograph of an artery in the connective tissue of the prosobranch *Littorina littorea*. E, endothelial cells; M, muscle cells intermingled with collagenous connective tissue fibres; V, vesicular connective tissue cell. Azan stain. × 400.

III. Structure and classification of blood cells

In spite of the fact that numerous studies have been conducted on blood
cells of gastropods, there is still no agreement on the question of how many
cell types can be distinguished (e.g. George and Ferguson, 1950; Müller,
1956; Kostal, 1969; Cheney, 1971; Sminia, 1972; Cheng and Auld, 1977).
The disagreement originates from the use of different "cell type" concepts.
Many investigators consider cells with slight morphological differences as
different "cell types" (e.g. Müller, 1956; Kress, 1968; Cheng et al., 1969),
while others stress the importance of functional differences (e.g. Brown and
Brown, 1965; Prowse and Tait, 1969; Davies and Partridge, 1972).
Moreover, the terminology employed for the different cell types is often
unclear. The adoption of terms used in vertebrate haematology, such as
leucocyte, granulocyte, macrophage and lymphocyte, has created much
unnecessary confusion. In this section an attempt will be made to re-classify
the blood cells of gastropods and a common nomenclature will be proposed.

Blood cells have been studied in about 25 species of gastropods. Table I
shows that 1–4 "types" of cell were distinguished by authors who only used
light microscopy for their studies (first 12 refs). Those who used also
electron microscopy distinguished 1–2 types (refs 13–19). The main criteria
used for the classification of blood cells at the light microscope level are: the
size of the cell, the nucleus to cytoplasm ratio, the shape of the nucleus, the
ability to form pseudopodia, and the presence or absence of cytoplasmic
granules.

On comparing the light microscopical descriptions it seems possible to
distinguish two classes: (1) round, relatively small cells with a high nucleus
to cytoplasm ratio (e.g. lymphoid cells of *Busycon* species, George and
Ferguson, 1950; normal cells of *L. stagnalis*, Müller, 1956; small agranular
leucocytes of *L. scabra*, Cheng et al., 1969), and (2) spreading cells with
large pseudopodia, a polymorphic nucleus and cytoplasmic granules (e.g.
type B cell of *H. aspersa*, Wagge, 1955; granular macrophages of *Bullia*
species, Brown and Brown, 1965; large granular leucocytes of *L. scabra*,
Cheng et al., 1969). It appears that cells of class 2 vary considerably more in
morphology than those of class 1. It has been reported that the majority of the
cells in the haemolymph (about 95 % or more) are spreading cells (e.g. George
and Ferguson, 1950; Brown and Brown, 1965; Davies and Partridge, 1972).
Because intermediate forms between the distinguished cell "types" have
been observed frequently (Table I), it seems unlikely that these "types"
represent fully differentiated cells. According to Brown and Brown (1965),
the round cells constitute a population of young cells which still have the
capacity to divide (see also section on haemocytopoiesis). The spreading
cells, on the other hand, are supposed to be mature cells which are active in

TABLE I. Nomenclature and morphological characteristics of gastropod blood cells.

Species	General indication of blood cells	Cell types distinguished	Morphology	Size (μm)	Proposed new nomenclature	Reference
12 species (Pulmonata)	amoebocytes	1 type: amoebocyte	with pseudopodia; granular cytoplasm	—	spreading amoebocyte	1. Cuénot (1892)
22 species (Pulmonata, Prosobranchia, Opisthobranchia)	leucocytes	2 types: 1. stage I cell 2. stage II cell	1. small, round, hyaline cell with spherical nucleus 2. large cell with polymorphic nucleus (intermediate stages present)	— —	hyalinocyte spreading amoebocyte	2. Kollmann (1908)
Helix aspersa (Pulmonata)	amoebocytes	1 type: amoebocyte	large cell, with polymorphic nucleus and pseudopodia	—	spreading amoebocyte	3. Baecker (1932) Haughton (1934)
Busycon carica B. canaliculatum Fasciolaria tulipa (Prosobranchia)	blood cells	3 types: 1. lymphoid cell 2. granular macrophage 3. eosinophilic granular amoebocyte	1. small, round, hyaline cell 2. large cell with granules; polymorphic nucleus 3. large cell with eosinophilic granules (intermediate stages present)	— — —	hyalinocyte spreading amoebocyte spreading amoebocyte	4. George and Ferguson (1950)
Helix aspersa (Pulmonata)	amoebocytes	2 types: 1. Type A 2. Type B	1. small, hyaline cell with spherical nucleus 2. large granular cell	10–20 up to 50	hyalinocyte spreading amoebocyte	5. Wagge (1955)

TABLE I.—*cont.*

Species	General indication of blood cells	Cell types distinguished	Morphology	Size (μm)	Proposed new nomenclature	Reference
Lymnaea stagnalis (Pulmonata)	leucocytes	4 types: 1. normal cell 2. leucocyte 3. small cell 4. wandering cell	1. small, round cell 2. actively amoeboid phase of normal cell 3. possibly an artefact 4. phagocytic normal cell (intermediate stages present)	10–15	round amoebocyte spreading amoebocyte ? spreading amoebocyte	6. Müller (1956)
Australorbis glabratus (Pulmonata)	amoebocytes	1 type: amoebocyte	cell with pseudopodia and granular cytoplasm; polymorphic nucleus	9–12	spreading amoebocyte	7. Pan (1958)
Bullia laevissima *B. digitalis* (Prosobranchia)	haemocytes	2 types: 1. lymphoid cell 2. granular macrophage	granular macrophages are more numerous and larger than lymphoid cells (intermediate stages present)	—	round amoebocyte spreading amoebocyte	8. Brown and Brown (1965)
Doto fragilis *D. coronata* *D. pinnatifida* (Opisthobranchia)	leucocytes	3 types: 1. lymphocyte I 2. lymphocyte II 3. amoebocyte	1. small round cell 2. large phagocytic cell 3. granular cell (intermediate stages present)	4–6 6–15 —	hyalinocyte spreading amoebocyte spreading amoebocyte	9. Kress (1968)
Littorina scabra (Prosobranchia)	leucocytes	3 types: 1. small agranular leucocytes 2. small granular leucocytes 3. large granular leucocytes	1. small, hyaline, round cell with spherical nucleus 2. small cell with granules, bilobed nucleus and pseudopodia 3. large granular cell with pseudopodia and sub-spherical nucleus	4 4–5 6–7	hyalinocyte spreading amoebocyte spreading amoebocyte	10. Cheng *et al.* (1969)

Species	Term	Types	Description	Size/Number	Functional form	Reference
Patella vulgata (Prosobranchia)	haemocytes	2 types: 1. macrophage 2. amoebocyte	1. round cell with granules 2. large cell with pseudopodia	30–50 10–20 × 25–35	? spreading amoebocyte	11. Davies and Partridge (1972)
Biomphalaria glabrata (Pulmonata)	haemocytes	2 types: 1. hyalinocyte 2. granulocyte	1. small, hyaline, round, cell without pseudopodia 2. large granular cell with pseudopodia	4 *ca.* 36	hyalinocyte spreading amoebocyte	12. Cheng and Auld (1977)
Lymnaea stagnalis (Pulmonata)	amoebocytes	1 type: amoebocyte (occurring in 2 functional forms?)	cell with pseudopodia, polymorphic nucleus large morphological variation from round to spread	*ca.* 10	(round and spreading) amoebocytes	13. Stang-Voss (1970)
Lymnaea stagnalis (Pulmonata)	amoebocytes	1 type: amoebocyte	round to spreading cell forms. Large morphological variation. Spherical to polymorphic nucleus	10–70	(round and spreading) amoebocytes	14. Sminia (1972)
Biomphalaria pfeifferi (Pulmonata)	amoebocytes	1 type: amoebocytes	round to spreading cells with pseudopodia. Cytoplasmic granules present	7	(round and spreading) amoebocytes	15. Meuleman (1972)
Biomphalaria glabrata (Pulmonata)	haemocyte	1 type: haemocyt	large cell with pseudopodia, cytoplasmic granules; polymorphic nucleus	—	spreading amoebocytes	16. Faulk *et al.* (1973)
Biomphalaria glabrata (Pulmonata)	haemolymph cells	2 types: 1. hyalinocyte 2. granulocyte	1. agranular, small round cell 2. large cell with pseudopodia and cytoplasmic granules	*ca.* 7 *ca.* 29	hyalinocyte spreading amoebocyte	17. Cheng (1975)
Biomphalaria glabrata (Pulmonata)	haemolymph cells	2 types: 1. hyalinocyte 2. granulocyte	1. agranular, hyaline cell 2. large cell with pseudopodia and cytoplasmic granules	— —	hyalinocyte spreading amoebocyte	18. Harris (1975)
Cerithidea californica (Prosobranchia)	haemolymph cells	2 types: 1. hyalinocyte 2. granulocyte	1. agranular, hyaline, round cell 2. granular cytoplasm with pseudopodia	4–8 9–15	hyalinocyte spreading amoebocyte	19. Yoshino (1976)

phagocytosis, encapsulation and wound healing (see below). Proof for the hypothesis that the distinguished cell types represent different developmental stages of one basic cell type has, however, not been provided.

Ultrastructural and histochemical studies on gastropod blood cells have also failed to solve the question of the number of blood cell types. A number of investigators, all working primarily on different aspects of trematode-snail relationships, are of the opinion that gastropods possess two distinct types of blood cell, granulocytes and hyalinocytes (Harris, 1975; Yoshino, 1976; Cheng and Auld, 1977; Krupa *et al.*, 1977). On the other hand, Stang-Voss (1970) and Sminia (1972), who both studied *L. stagnalis*, think that there is only one type of blood cell, the amoebocyte.

The morphology of the granulocytes, hyalinocytes and amoebocytes will be described successively. The descriptions of granulocytes and hyalinocytes are based on studies of blood cells in the prosobranch *Cerithidae californica* (Yoshino, 1976) and in the pulmonates *Biomphalaria glabrata* (Cheng, 1975; Harris, 1975; Cheng and Auld, 1977; LoVerde, personal communication), *Bulinus guernei* (Krupa *et al.*, 1977) and in *Bulinus alexandrina*, *B. truncatus*, *B. tropicus* and *B. natalensis* (LoVerde, personal communication). The description of amoebocytes is based on work on *L. stagnalis* (Stang-Voss, 1970; Sminia, 1972, 1977a).

A. Granulocytes

Figures 4 and 6 show that granular blood cells when allowed to settle on glass show many pseudopodia with ribs and spikes (Cheng, 1975). The pseudopodia have a hyaline appearance and the majority of the cell organelles are located in a rim around the usually polymorphic nucleus (Fig. 4). The dimensions of the granulocytes seem to vary between species and are largely dependent on the preparation and observation methods used. Diameters between 9 and 70 μm have been reported (Yoshino, 1976; Cheng and Auld, 1977; LoVerde, personal communication). In blood smears, the nucleus is deeply-stained with routine methods, the rim of perinuclear cytoplasm is slightly basophilic and the pseudopodia remain clear. Ultrastructural examinations of the granulocytes reveal that they have numerous mitochondria, smooth and rough endoplasmic reticulum, Golgi bodies, glycogen deposits, lipid droplets and lysosome-like structures (vacuoles containing whorls of membranes and electron–dense material) (Fig. 6). The extensiveness of the endoplasmic reticulum varies considerably from cell to cell as do the number and the development of the other cell organelles and inclusions. Enzyme histochemical studies have shown the presence of acid phosphatase positive granules in granulocytes (Harris and

Cheng, 1975a; Cheng and Garrabrant, 1977). Moreover it appears that the lysosome-like structures and the Golgi apparatus have a high peroxidase activity (Carter and Bogitsh, 1975).

B. Hyalinocytes

Hyalinocytes are smaller than granulocytes (Figs 5, 7). They are less than 8 μm in size (Yoshino, 1976; LoVerde, personal communication). Moreover, the nucleus to cytoplasm ratio is higher. Like granulocytes the hyalinocytes adhere to glass, but they remain spherical (Fig. 5) and do not form large pseudopodia with ribs or spikes. Usually the nucleus is round and the fine granular cytoplasm is uniformly basophilic. The cells are only slightly stained with routine blood stains. Also, in the electron microscope, the cells are electron-transparent (Fig. 7). The cytoplasm contains mitochondria, Golgi bodies, rough endoplasmic reticulum, and vesicles of various size and electron densities. The lysosomal enzyme acid phosphatase, present in granulocytes, could not be detected in hyalinocytes (Harris and Cheng, 1975a; Cheng and Garrabrant, 1977). Nothing is known about the presence or absence of peroxidase in hyalinocytes.

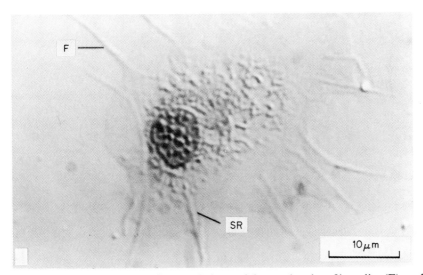

Fig. 4. Spread granulocyte of *Biomphalaria glabrata* showing filopodia (F) and supporting "rods" (SR, spikes). (Nomarski interference optics). [Courtesy of Dr T. C. Cheng.]

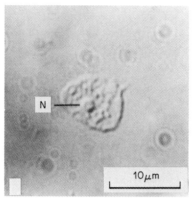

Fig. 5. Spread hyalinocyte with conspicuous nucleus (N) of *B. glabrata* (Nomarski interference optics). [Courtesy of Dr T. C. Cheng.]

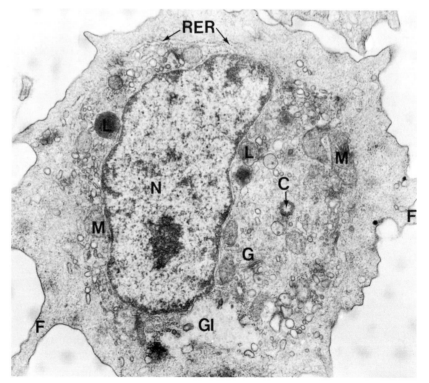

Fig. 6. Granulocyte of *Biomphalaria glabrata*. C, centriole; F, filopodia; G, Golgi apparatus; Gl, glycogen; L, lysosome; M, mitochondrion; N, nucleus; RER, rough endoplasmic reticulum. × 15 000.

C. Amoebocytes

Immediately following haemolymph withdrawal amoebocytes (Figs 8–12) become adhesive and they rapidly begin to aggregate. In suspension, the cells are small and round (diameter about 8 μm), but when in contact with glass the majority flatten and spread over the surface by extending pseudopodia with spikes (Fig. 8). The cells can attain lengths of 70 μm. A few cells maintain their round shape; they do not form pseudopodia and spread only slightly (Fig. 8). Cells which can be considered as intermediate stages between spreading and round amoebocytes are always present. The shape of the nucleus of spreading amoebocytes varies considerably, from round or oval, to kidney-shaped and lobulated. The cytoplasm is rather vacuolated and contains granules. As these are acid phosphatase and peroxidase positive they are considered to be lysosomes. Their number varies from cell to cell. In stained blood smears, amoebocytes have an appearance similar to that of granulocytes. Round amoebocytes are more basophilic than the spreading ones. Ultrastructural studies on pellets of

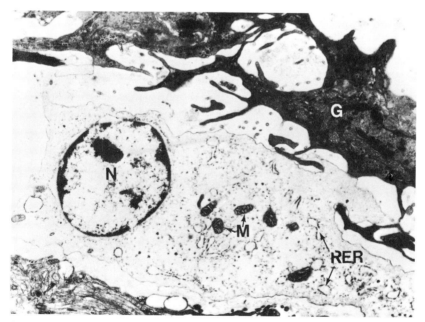

Fig. 7. Electron micrograph of hyalinocyte of *Bulinus tropicus*, showing the electron-lucent appearance of this cell type. G, granulocyte; M, mitochondrion; N, nucleus; RER, rough endoplasmic reticulum. [Courtesy of Dr P. T. LoVerde.] × 6000.

amoebocytes as well as on amoebocytes fixed *in situ*, within the circulatory system, also reveal a great morphological variation with respect to the extent of the rough endoplasmic reticulum. In some amoebocytes, part of the rough endoplasmic reticulum is enormously expanded and contains fine granular material (cf. Stang-Voss, 1970; Sminia, 1972). The number and appearance of lysosomes, the development of the Golgi system, the amount of glycogen and the shape of the nucleus also show considerable variation (compare Figs 9, 10). It is evident from these observations that the amoebocytes form a heterogeneous population of cells. Their main features are the presence of: (1) lysosomes with fine granular or crystalline-like contents (Sminia, 1972; Fig. 11), (2) filamentous pseudopodia (filopodia; Fig. 10), and (3) peroxidases in both the Golgi system and the lysosomes (Fig. 12). Compared to the spreading amoebocytes, the round amoebocytes have more extensive rough endoplasmic reticulum and a greater number of free ribosomes (Figs 9, 10). The cytoplasm of these cells is also more electron-dense. Similar observations have been made on round amoebocytes from *Helix aspersa* (Tait, personal communication).

On the basis of these observations, it is difficult to decide whether round and spreading amoebocytes should be considered as one or as two cell types. Whereas on the one hand, the presence of morphologically intermediate stages between round and spreading amoebocytes indicates that the

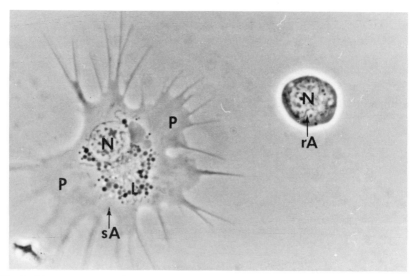

Fig. 8. Phase contrast micrograph of a round (rA) and a spreading (sA) amoebocyte of *L. stagnalis*. L, lysosomes; N, nucleus; P, pseudopodia with spikes. × 1500.

cells concerned belong to the same cell type, their behaviour on glass is different, which favours the two-cell-type theory. On comparing spreading amoebocytes with granulocytes it seems clear that they should be considered as one cell type. They are similar in morphology and cytochemistry as well as in their behaviour on glass. Furthermore, they show great similarity in functional capacities (see below).

The hyalinocytes, finally, differ from the granulocytes (amoebocytes) as well as from round amoebocytes; from the latter in morphological aspects, from the former also in behaviour. They thus form a separate cell type.

It should be stressed that until now, only morphological and behavioural criteria have been used. The question of the number of blood cell types must be further considered in the light of studies on the functions and the development of the cells. Unfortunately, however, the controversy concern-

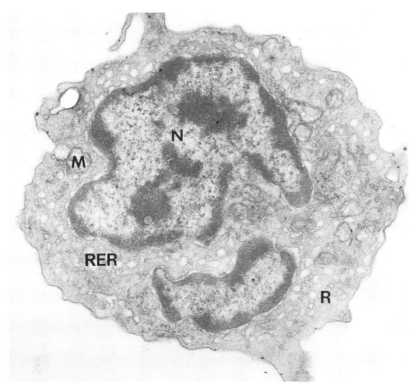

Fig. 9. Electron micrograph of a round amoebocyte of *L. stagnalis*. Note the rough endoplasmic reticulum (RER) and the numerous ribosomes (R). M, mitochondrion; N, nucleus. × 19 000.

ing round and spreading amoebocytes still remains. Prowse and Tait (1969),
reported that round and spreading amoebocytes of *H. pomatia* also differ in
functional aspects: round amoebocytes do not phagocytose, whereas
spreading cells are highly phagocytic. In *L. stagnalis*, on the other hand,
round amoebocytes do phagocytose, although to a much lesser degree than
spreading cells, which probably indicates that in this snail only one type of
blood cell is present (Sminia, unpublished). Investigations on blood cell
formation have also failed to clarify whether or not the round and spreading
amoebocytes are distinct cell types (see section on haemocytopoiesis).

Finally, although extensive studies on the functions of hyalinocytes are

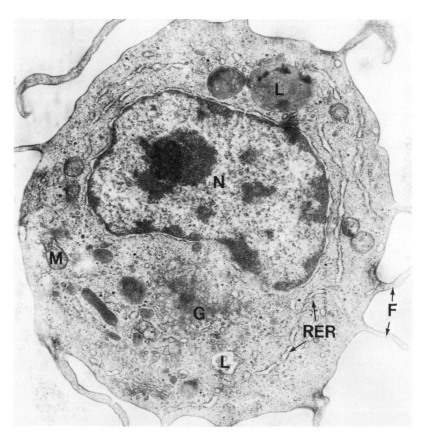

Fig. 10. Electron micrograph of a spreading amoebocyte of *L. stagnalis*, fixed
immediately following haemolymph withdrawal. F, filopodia; G, Golgi apparatus;
L, lysosome; M, mitochondrion; N, nucleus; RER, rough endoplasmic reticulum.
× 15 000.

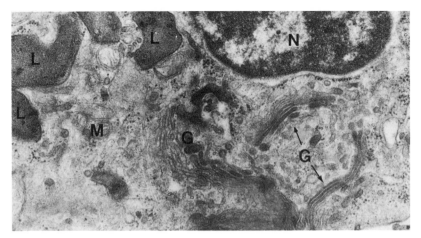

Fig. 11. Detail of the Golgi apparatus (G) of a spreading amoebocyte of *L. stagnalis* showing the formation of lysosomes (L). M, mitochondrion; N, nucleus. × 27 000.

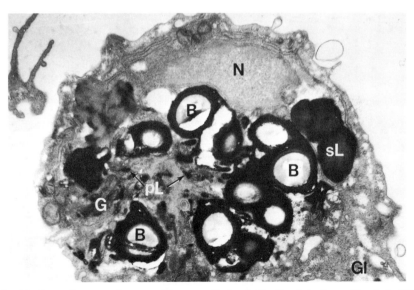

Fig. 12. Spreading amoebocyte of *L. stagnalis* fixed 4 h after injection of bacteria (B) and incubated for the demonstration of peroxidases. The black reaction product is present in the Golgi apparatus (G), primary (pL) and secondary lysosomes (sL). Gl, glycogen; N, nucleus. × 10 000.

lacking, preliminary observations on phagocytosis and encapsulation of foreign material in *B. glabrata* (Cheng and Garrabrant, 1977) suggest that hyalinocytes are not involved in these processes, while amoebocytes (granulocytes) are.

D. Nomenclature

In recent years, both terms, amoebocytes and granulocytes, have been used for gastropod blood cells. The term granulocyte strongly suggests that these cells are rather similar to vertebrate granulocytes. Although both are phagocytic and contain peroxidase, there are several differences between gastropod and vertebrate granulocytes. For example, vertebrate granulocytes are fully differentiated cells, filled with granules and unable to divide. Gastropod granulocytes, on the other hand, often contain only a few granules, are able to form new ones (lysosomes), and can still divide; their pseudopodia (with spikes) are also different from those of vertebrate granulocytes. In fact, gastropod granulocytes much more resemble vertebrate monocytes (macrophages; see also Chapter 17). Therefore, to avoid any confusion, I propose to dispense with vertebrate terms, and to use the neutral term amoebocyte. Finally, until more is known about their function the neutral term hyalinocyte can be used for the electron-transparent cells occurring in gastropod haemolymph.

IV. Origin and formation of blood cells

A. Haemocytopoiesis

In a number of morphological studies on gastropods, reference has been made to blood cell formation. Certain epithelia and various parts of the connective tissue have been regarded as proliferation sites of blood cells: the epithelium and the connective tissue of the mantle (*H. aspersa*, Wagge, 1951, 1955), the connective tissue around the lung cavity (*L. stagnalis*, Müller, 1956), the walls of blood sinuses and the connective tissue in the kidney region (*B. glabrata*, Pan, 1958), the ventricle and the connective tissue around the heart-kidney system (*Doto*-species, Kress, 1968) and the connective tissue of the mantle (*Bulinus*-species, Kinoti, 1971). It has been suggested that the blood cells originate: (1) from precursor cells ("embryonic cells", Müller, 1956; "amoeboblasts", Kinoti, 1971), (2) from mature amoebocytes (Cuénot, 1892; George and Ferguson, 1950; Brown and Brown, 1965), or (3) from fibroblasts, epithelial or endothelial cells (Wagge, 1951, 1955; Pan, 1958). Three mechanisms of proliferation have been mentioned, viz. mitosis, amitosis and cytoplasmic fragmentation (e.g. Cuénot, 1892; Wagge, 1951).

This brief summary shows that there is no consensus on the following main aspects of haemocytopoiesis, (1) the sites of proliferation, (2) the mechanism of proliferation and (3) the question as to whether or not stem cells are present. These three points have been the subject of two recent papers (Sminia, 1974; Lie et al., 1975).

According to Sminia (1974), who studied haemocytopoiesis by use of electron microscopical and autoradiographical techniques, L. stagnalis does not possess special haemocytopoietic organs, as blood cell proliferation takes place throughout the body. This conclusion is based on the following observations: (1) dividing amoebocytes occur in both the haemolymph and the connective tissue, (2) haemolymph as well as connective tissue amoebocytes take up ³H-thymidine and thus synthesize DNA (autoradiography; Fig. 13), (3) the percentage of thymidine-labelled amoebocytes is the same in all

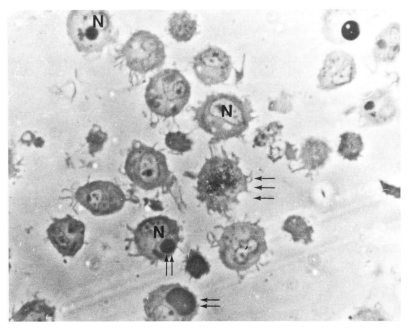

Fig. 13. Autoradiograph of amoebocytes of L. stagnalis injected with ³H-thymidine 1 h prior to haemolymph sampling. Note that in one of the cells (3 arrows) the nucleus is labelled, indicating that circulating blood cells are able to divide. The large black granules (double arrows) represent dilated parts of the rough endoplasmic reticulum. N, nucleus. (1 μm thick Epon section of blood cell pellet). × 2000.

body parts. As the heart-kidney region has a very high number of amoebocytes (Sminia, 1972) this body part is the main production site of amoebocytes. This conclusion is supported by observations on blood cell formation in *B. glabrata*, infected and reinfected with echinostome trematodes (Lie *et al.*, 1975; Lie and Heyneman, 1976). Although it was first reported that in this snail amoebocytes are exclusively formed in the connective tissue between the pericardium and the posterior epithelium of the mantle cavity ("amoebocyte-producing organ", Fig. 14; Lie *et al.*, 1975), later studies (e.g. Lie and Heyneman, 1976) mention that proliferation of amoebocytes also takes place in other parts of the body.

With respect to the mechanisms of proliferation, the results of these recent studies have clearly shown that amoebocytes proliferate by mitosis.

There is to date no certainty as to what type of cells gives rise to blood cells; are stem cells present or not? It has been shown that circulating amoebocytes can divide. Moreover, amoebocytes involved in phagocytosis and encapsulation of foreign material are still able to incorporate ^{3}H-thymidine (Sminia, 1974). Although these observations indicate that mature amoebocytes can generate daughter cells, the presence of a special pool of stem or embryonic cells, as suggested by several authors, cannot be ruled out (e.g. Müller, 1956; Kinoti, 1971; Lie *et al.*, 1975). Brown and Brown (1965), thought that the round amoebocytes constitute such a population of dividing cells. The fact that round amoebocytes have the morphological features of young cells is in line with this hypothesis.

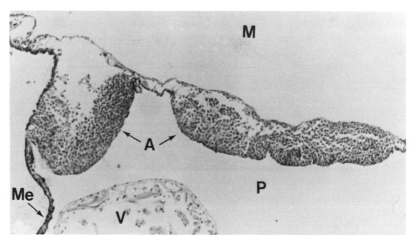

Fig. 14. Section through the amoebocyte producing organ (A) of the freshwater snail *Biomphalaria glabrata*. M, mantle cavity; Me, mantle epithelium; P, pericardial cavity; V, ventricle. Azan stain. × 250.

B. The number of haemocytes

The density of blood cells in the haemolymph has been established for several gastropods (Table II). Counts were carried out on samples of haemolymph taken from several body parts, primarily from the heart and the foot region. As can be seen in Table II the number of cells per ml of haemolymph varies from species to species and within a species shows a large variation. The average number of cells per ml is about 5×10^5.

TABLE II. Numbers of circulating blood cells of several gastropod species.

Species	Number of blood cells per ml haemolymph	Reference
Bullia laevissima	$3·8 - 7·2 \times 10^6$	Brown and Brown (1965)
Helix pomatia	$0·2 \times 10^6$	Bayne (1974)
Patella vulgata	$1·0 - 9 \times 10^6$	Davies and Partridge (1972)
Lymnaea stagnalis	$0·5 - 4 \times 10^6$	Sminia (1972)
Biomphalaria glabrata	$0·1 - 1 \times 10^6$	Jeong and Heyneman (1976), Cheng and Auld (1977), Stumpf and Gilbertson (1978)

However, the counts should be considered with great caution since the method used to extract haemolymph markedly influences the number of cells (Sminia, unpublished). There can also be a considerable difference between samples of haemolymph taken from different sites. For example, in *Bullia laevissima*, haemolymph samples from the heart and arteries are richer in blood cells than those taken from the veins, while the sinuses show a still lower blood cell number (Brown and Brown, 1965). Moreover, the number of circulating cells will also probably differ in snails living under different conditions.

A few examples of possible causes of the large variation in the number of blood cells are given. The number of circulating blood cells in *B. glabrata* (Stumpf and Gilbertson, 1978) and in *L. stagnalis* (Sminia, unpublished, Fig. 15) seems to be related to the age of the animals. Large specimens (30 mm shell height) have about 2–3 times as many cells per unit volume of haemolymph than small ones (10 mm). The results of Davies and Partridge (1972) point to a temperature dependence: in *Patella vulgata* the concentration of circulating blood cells varies from about 1×10^6 cells per ml at $5°C$ to 9×10^6 cells per ml at $25°C$. Also, in *B. glabrata* and *L. stagnalis*, the number of circulating cells increases rapidly (within a few hours) when the temperature rises (Stumpf and Gilbertson, 1978; Sminia, unpublished). Other factors influencing the number of blood cells are infection, wounding and haemolymph extraction. Immediately after injection of foreign particles the number of circulating blood cells of snails decreases (*H. pomatia*, Bayne, 1974; *L. stagnalis*, van der Knaap, personal communication). After this initial decrease, in *B. laevissima*, the number increases within 7 days to about 2–5 times the original number (Brown and Brown, 1965). In *L. stagnalis* the number of blood cells also increases considerably but within a much shorter period (1 h) after incision or haemolymph withdrawal (Müller, 1956; Sminia, 1972). As the increase in the number of circulating amoebocytes is very rapid and as no increase in the number of mitotic figures was found, under these circumstances, it is unlikely that the increase is due to cell division. More probably, the rapid changes in the number of blood cells are due to release of cells from the connective tissue into the circulatory system. The connective tissue around the heart-kidney region and the central connective tissue (Bekius, 1972) may serve as a reservoir of blood cells, which under certain conditions, are emptied into the circulation (Sminia, 1974). The function of this redistribution of blood cells may be interpreted in terms of enhancing the effectiveness of the cellular defence reactions.

V. Functions of blood cells

A. Phagocytosis

Since Tripp's (1961) work on the fate of foreign particles injected into the freshwater snail, *Australorbis glabratus*, several studies have been made on the role of amoebocytes in the internal defence of gastropod molluscs (see reviews by Tripp, 1970; Bang, 1975; Michelson, 1975; Cooper, 1976; Anderson, 1977). Numerous types of foreign particles, biotic as well as abiotic (e.g. ferritin, viruses, bacteria, red blood cells, yeast, India ink, colloidal suspensions of particles of various sizes) have been used. From such studies it has become evident that amoebocytes are highly phagocytic (Figs 12, 16). In gastropods, phagocytosis is a major line of defence against invading microorganisms.

The initial steps in phagocytosis, i.e. the first interaction between amoebocytes and foreign particles and the subsequent adhesion, have not been studied in detail in gastropods. Thus relatively little is known on the question whether the initial interaction comes about by chance or whether amoebocytes are chemotactically attracted by the foreign material. Schmid (1975) has shown, in an *in vitro* system, that amoebocytes of the snail *Viviparus malleatus*, are chemotactically attracted by heat-killed

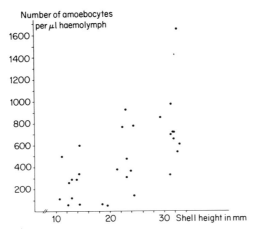

Fig. 15. Relationship between the number of amoebocytes per unit volume of haemolymph and shell height for *L. stagnalis*.

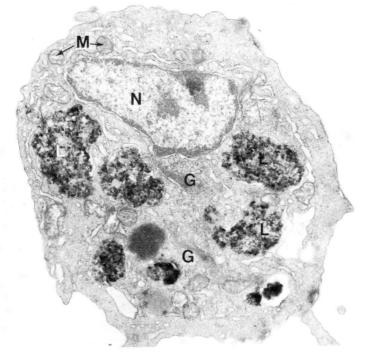

Fig. 16. Electron micrograph of a spreading amoebocyte of *L. stagnalis* fixed 24 h after an India ink injection. G, Golgi apparatus; L, lysosome containing ink particles; M, mitochondria; N, nucleus. × 9000.

Staphylococcus aureus bacteria and by *N*-acetyl-D-glucosamine. However, chemotaxis only occurred in the presence of certain soluble plasma constituents which possess bacterial agglutinating properties. Whether or not this is a general phenomenon among gastropods has to be established by studying amoebocyte-foreign body interactions in other species. Relative to the adhesion of foreign particles to amoebocytes, it has been established that this process as well as phagocytosis of the foreign particles are facilitated by serum factors (see below).

Morphological and ultrastructural studies have shown that amoebocytes phagocytose foreign particles by (1) extending pseudopodia which engulf the foreign particles, or (2) by forming invaginations of the plasma membrane. Both processes result in the particles becoming enclosed in phagosomes and digested in the lysosomal system. The phagosomes fuse with primary and secondary lysosomes in which the hydrolyzing enzymes acid phosphatase (*L. stagnalis*, Sminia, 1972; Sminia *et al.*, 1974; *B. pfeifferi*, Meuleman, 1972; *B. glabrata*, Carter and Bogitsh, 1975) and non-specific esterases (*B. glabrata*, Harris, 1975; Harris and Cheng, 1975a) have been demonstrated. Furthermore, peroxidases are present in the lysosomal system of the amoebocytes of *L. stagnalis* and *B. glabrata* (Fig. 12; Sminia, 1972; Carter and Bogitsh, 1975). It has been suggested that the peroxidases of the amoebocytes function similarly to those in mammalian leucocytes, i.e. as bactericidal, viricidal and fungicidal agents (Sminia, 1972). Peroxidase activity has not only been observed within lysosomes, but also in the Golgi system (Fig. 12). In amoebocytes that have phagocytosed foreign particles the number of lysosomes is increased. Furthermore, histochemically, acid phosphatase activity was more apparent in these cells than in non-phagocytosing amoebocytes (Sminia, 1972). These observations indicate that phagocytosis induces the activity and production of this lysosomal enzyme.

At present, nothing is known about the biochemical aspects of phagocytosis in gastropod amoebocytes. It has been suggested that lysosomal enzymes are released into the serum where they act as humoral defence agents (e.g. Cheng *et al.*, 1977, 1978). However, conclusive evidence for this suggestion has not been presented.

Several reports indicate that gastropods eliminate foreign material by means of a special mechanism in which amoebocytes laden with phagocytosed foreign particles leave the animal by active migration to the external environment, via the epithelia of the kidney, mantle, gut, reproductive tract and foot (e.g. Tripp, 1961; Brown and Brown, 1965; Brown, 1967; Cheng *et al.*, 1969). This manner of elimination of foreign material is, however, not a general phenomenon among gastropods. For example, in *L. stagnalis* and *B. glabrata*, amoebocytes laden with foreign particles are still found in the haemolymph and in the connective tissue three months after injection of the particles (Tripp, 1961; Sminia, 1972).

Haemolymph amoebocytes often settle in the connective tissue of various body parts, temporarily or more permanently, singly or in groups and especially in the heart-kidney region. These tissue amoebocytes, which in *L. stagnalis* are morphologically similar to haemolymph amoebocytes (Sminia, 1974), can also ingest foreign particles. In the connective tissue of *H. pomatia* a second phagocytosing cell type, a fixed phagocyte, has been described (Reade, 1968; Bayne, 1973). This cell type is mainly found in the digestive gland region and around blood vessels. Recently Sminia *et al.* (1979b) demonstrated that in *L. stagnalis* fixed phagocytic cells also occur throughout the connective tissue (Fig. 17). These cells differ ultrastructurally as well as histochemically (they do not possess peroxidases) from amoebocytes, which suggests that the two cell types also differ functionally.

In addition to these two types of highly phagocytic cells, gastropods possess another cell type which is capable of endocytosis This cell type, the pore cell, occurs throughout the connective tissue (e.g. Buchholz *et al.*, 1971; Sminia, 1972). Experiments in which snails were injected with particles of different nature and size have shown that pore cells have a high

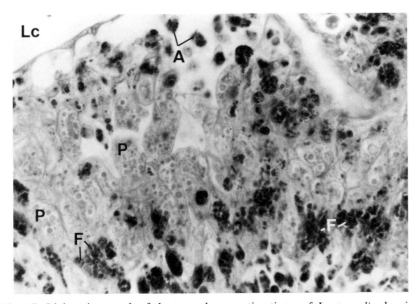

Fig. 17. Light micrograph of the central connective tissue of *L. stagnalis* showing bacteria-laden free (A) and fixed (F) phagocytic cells. Lc, lung cavity; P, pore cell. Alcian blue eosin stain. × 560.

affinity for proteinaceous materials and, furthermore, they only ingest particles with a diameter of less than 20 nm (Wolburg-Buchholz and Nolte, 1973; Boer and Sminia, 1976; Sminia, 1977b). The functional significance of this selective endocytosis is still unknown. As mentioned above, the main function of pore cells is the synthesis, storage and release of blood pigment (e.g. Sminia, 1977b).

Two other lines of approach, viz. clearance and *in vitro* investigations, have been used to study the role of amoebocytes in the removal and degradation of harmful foreign substances.

1. Clearance studies

Bayne and Kime (1970) demonstrated that *H. pomatia* eliminates bacteria very rapidly (about 90% within 2 h) from its haemolymph, even if very large numbers are injected. Similar observations were made on *L. stagnalis* (Fig. 18; van der Knaap *et al.*, in preparation). According to Bayne (1974) both the haemolymph amoebocytes and the fixed phagocytes are responsible for the removal of bacteria. Recent morphological studies on *L. stagnalis* injected with large numbers of live bacteria (*Escherichia coli* and

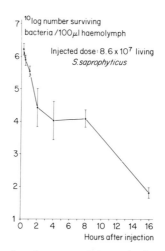

Fig. 18. Graph showing the clearance of bacteria from the haemolymph of *L. stagnalis*. At zero time 35 adult specimens were injected with 8.6×10^7 living *Staphylococcus saprophyticus*; haemolymph samples were taken at 7 consecutive points of time (5 snails per point). The number of bacteria present at the moment of sampling was determined using a microbiological technique (spread plate method). Mean and standard deviation of the samples were plotted semilogarithmically. [Courtesy of W. P. W. van der Knaap.]

Staphylococcus saprophyticus) have confirmed this finding (Sminia *et al.*, 1979b; van der Knaap *et al.*, in preparation). Much work, however, has still to be done to unravel the role of the different types of phagocytes in the removal and degradation of microorganisms.

Although it has been suggested that in addition to the phagocytes, humoral factors such as bactericidins, lysins and agglutinins may be involved in the clearance of microorganisms from the haemolymph, evidence for such factors is still lacking (e.g. Pauley *et al.*, 1971).

2. In vitro *studies*

The first *in vitro* studies on the role of serum factors in phagocytosis were carried out in *H. aspersa* by Prowse and Tait (1969). These authors stated that serum factors—called opsonins as they coat the foreign particles—are indispensible for phagocytosis of yeast cells and sheep erythrocytes by amoebocytes; the amoebocytes of this snail seem to lack plasma membrane receptors for the recognition of unopsonized foreign particles. In two other snails (*Otala lactea*, Anderson and Good, 1976; *L. stagnalis*, Sminia *et al.*, 1979a; Fig. 19), on the other hand, it has been shown that although

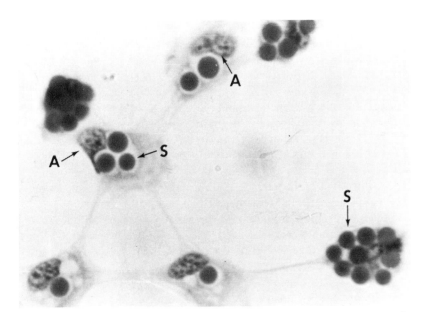

Fig. 19. Monolayer of amoebocytes (A) of *L. stagnalis* exposed for 1 h to sheep erythrocytes (S) in a snail serum-saline (1:1) medium. Haemalum-eosin stain. × 1500.

haemolymph opsonins enhance phagocytosis, they are not indispensible. This suggests that not only serum factors, but also plasma membrane receptors are involved in the recognition of foreign particles.

The presence of agglutinins which may have opsonic activity has been established in the serum of *Aplysia californica* (Pauley *et al.*, 1971) and *Viviparus malleatus* (Schmid, 1975), and in some other snails (Pemberton, 1974). Recently, Renwrantz and Cheng (1977a,b) demonstrated agglutinin binding receptors on the surface of *H. pomatia* amoebocytes.

Thus the observations strongly suggest that at least two kinds of receptors are present on the surface of amoebocytes: (1) aspecific receptors, which bind foreign particles without mediation of serum factors, and (2) receptors which can only bind particles coated with opsonins. As Prowse and Tait (1969) found a certain degree of specificity of the opsonic factors—yeast cells were not phagocytosed in serum absorbed with the homologous particles, whereas they were in serum absorbed with heterologous particles—it is not unlikely that different classes of specific receptors are present on the plasma membrane of amoebocytes.

B. Encapsulation

Studies on gastropod-parasite relationships have shown that sometimes defensive reactions against invading worm larvae occur. These are too large to be phagocytosed by single cells and it appears that they are encapsulated by large numbers of host cells (e.g. Faust, 1920; Pan, 1965; Cheng and Rifkin, 1970). The extent of the encapsulation reaction may differ greatly; it is dependent on the degree of compatibility of host and parasite (e.g. Sudds, 1960). In non-susceptible and in partially susceptible strains of *B. glabrata*, thick capsules are formed around the larvae of *Schistosoma mansoni* (Cheng and Garrabrant, 1977). In susceptible snails, on the other hand, viable larval stages (sporocysts and cercariae) are seldom encapsulated and only degenerating larvae of the parasite evoke a severe encapsulation reaction (e.g. Pan, 1965; Kinoti, 1971; Meuleman, 1972; Fig. 20).

Although several reports state that various types of host cells are involved in capsule formation (e.g. Tripp, 1961; Pan, 1965; Brooks, 1969; Cheng and Rifkin, 1970), recent electron microscope studies have shown that encapsulation is carried out exclusively by (granular) amoebocytes (Sminia *et al.*, 1974; Harris, 1975; Cheng and Garrabrant, 1977). Hyalinocytes do not seem to be involved in this process (Cheng and Garrabrant, 1977).

For the study of the formation and morphology of the capsules, gastropods were infected with parasites or with biologically inert objects. In other experiments, heterologous tissues were implanted into the snails (e.g.

Tripp, 1961; Sminia *et al.*, 1974). The results show that the encapsulation process is basically the same in all gastropods studied. Almost immediately after the introduction of a foreign object, the region in which it is located is infiltrated by amoebocytes. Within a few minutes the object is surrounded by a loose layer of amoebocytes. During the first three days, the number of amoebocytes involved in capsule formation increases continuously (Fig. 21; Sminia *et al.*, 1974; Cheng and Garrabrant, 1977). The amoebocytes which are directly attached to the foreign body flatten and form a continuous layer by extending processes into all irregularities of the surface of the implant. During the flattening process microtubules and bundles of micro-filaments appear in the peripheral cytoplasm of the cells. Apparently, these structures play an important role in cell elongation. From day 3 onwards, the capsule becomes stronger due to the fact that the pseudopodia of the amoebocytes interdigitate.

The further course of the encapsulation process seems to be influenced by the kind of parasite or foreign body that is encapsulated. According to

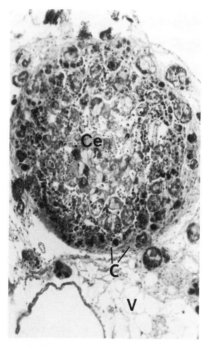

Fig. 20. Encapsulation of degenerating cercariae (Ce) in the freshwater snail *Biomphalaria pfeifferi*. V, vesicular connective tissue cells of the host. Amoebocyte capsule (C). Azan stain. [Courtesy of Dr E. A. Meuleman.] × 200.

Harris (1975), who worked with *B. glabrata* infected with first-stage larvae of the metastrongylid nematode, *Angiostrongylus cantorensis*, elongation of the amoebocytes also occurs in the outer parts of the capsule. As a result, the capsule finally consists of layer upon layer of thinly drawn out cells and pseudopodia. Neither extracellular connective tissue elements nor myofibroblasts (muscle cells which produce connective tissue elements) are found in these capsules.

On the other hand, in *L. stagnalis*, capsules formed around biologically inert implants only have the innermost cell layers flattened and closely packed (Sminia *et al.*, 1974). The initially flattened amoebocytes of the middle and especially of the outer capsule layers become separated by intercellular spaces filled with tiny connective tissue fibrils, in an amorphous ground substance (Fig. 22). Obviously, the intercellular substances are synthesized by amoebocytes which are transformed into fibroblasts. This assumption is based on the observations (1) that the cisternae of the rough endoplasmic reticulum become more dilated and extensive, indicating that

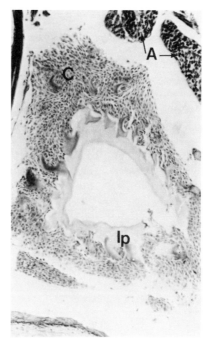

Fig. 21. Light micrograph of an encapsulated 24-h-old abiotic implant (Ip). A, tissue of the recipient snail (*L. stagnalis*); C, amoebocyte capsule. Azan stain. × 200.

the cells are actively synthesizing proteins, and (2) that the cells become labelled if ^{3}H-proline, a marker for collagen synthesis, is injected into the snails. In addition to amoebocytes and fibroblasts, myofibroblasts also appear in the capsule. Sminia *et al.* (1974) suggested that these cells, too, arise from amoebocytes.

The observations on host-parasite relationships suggest that gastropods are able to distinguish between viable and non-viable (degenerating) stages of the parasites. This discrimination capacity has also been shown in studies in which different types of tissue (autologous, homologous and heterologous) were implanted into snails (*B. glabrata*, Tripp, 1961; *Helisoma duryi normale*, Cheng and Galloway, 1970; *L. stagnalis*, Sminia *et al.*, 1974). It was found that autologous and homologous tissue implants are not encapsulated, whereas heterologous, as well as formalin-fixed and degenerated autologous tissues (Tripp, 1961; Sminia *et al.*, 1974), evoke a severe host reaction in which the implants are completely infiltrated and encapsulated within one day. The infiltrating amoebocytes phagocytose the degenerating implant. Eventually only a small globule ("granuloma"), mainly consisting of connective tissue, is left.

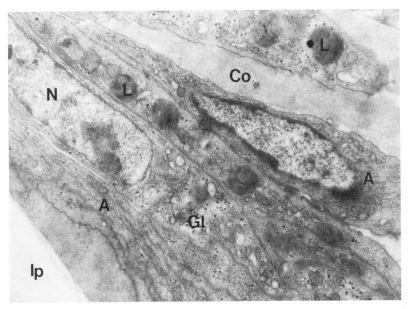

Fig. 22. Part of the capsule of a 3-week-old abiotic implant (Ip) in *L. stagnalis*, showing flattened amoebocytes (A) and collagenous connective tissue (Co). Gl, glycogen; L, lysosome; N, nucleus. × 12 000.

The fact that capsule amoebocytes phagocytose the implanted foreign tissue indicates that phagocytosing and encapsulating amoebocyes are the same cells. This conclusion is sustained by the fact that amoebocytes labelled with India ink (by injecting snails with ink 24 h before implantation) are involved in encapsulation (Sminia et al., 1974; Fig. 23).

The question of whether blood cells alone are responsible for the discrimination between different types of implant has not yet been studied. The studies on phagocytosis show that humoral factors are probably also involved.

C. Wound healing

The role of blood in the process of wound healing has only been studied in a small number of species, viz. in a few opisthobranchs (*Doto*-species, Kress, 1968), in one prosobranch, *Haliotis cracherodii* (Armstrong *et al.*, 1971) and in the pulmonate, *L. stagnalis* (Sminia *et al.*, 1973). These studies, of which the first two are light microscopical and the latter ultrastructural, showed that initial wound sealing is carried out by muscular contraction

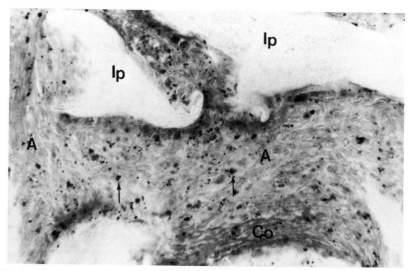

Fig. 23. Part of a capsule of a 14-day-old abiotic implant (Ip) in *L. stagnalis* which was injected with India ink 24 h before implantation. Note that the capsule forming amoebocytes (A) contain ink particles (arrows). Co, collagenous connective tissue. Azan stain. × 400.

and by the formation of plugs of blood cells; no coagulation of blood proteins was found to be involved.

The following account of wound healing is mainly based on the study of Sminia *et al.* (1973) on *L. stagnalis* in which short incisions were made in the skin. Almost immediately after wounding, the size of the wound opening is reduced by contraction of the muscles of the body wall. During the first 24 h, the wound region is infiltrated by amoebocytes. Those in the wound opening, aggregate and form a large plug which closes the wound completely within 9–24 h (Fig. 24). Extracellular tubules, most likely consisting of polymerized haemocyanin molecules, can be observed between the closely aggregated amoebocytes. At the margins of the wound, amoebocytes are very actively engaged in phagocytosing cell debris and injured tissue. Several structural changes occur in the cellular wound plug during the process of wound healing. Firstly, the amoebocytes become flattened; this process starts at the margins of the plug 18–24 h after wounding. Three to five days after incision, tiny collagen fibrils, embedded in an amorphous ground substance, appear between the flattened amoebocytes. As in the case of encapsulation, these transformed amoebocytes can be regarded as

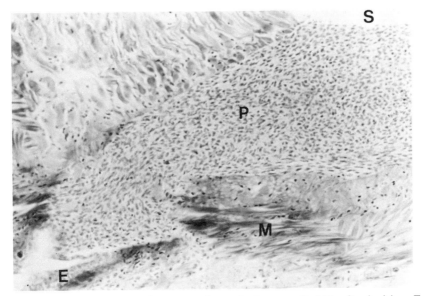

Fig. 24. Light micrograph of the wound area in *L. stagnalis* 24 h after incision. E, epidermis epithelium; M, muscle cells; P, amoebocyte plug in the wound; S, cephalopedal blood sinus. Azan stain. × 750.

fibroblasts, as they have the morphological characteristics of protein-producing cells and can incorporate large amounts of proline (as shown by autoradiography with the collagen marker ^{3}H-proline). From five days onwards, the amount of connective tissue ground substance and the number of fibrils increase. As a result of the formation of collagenous connective tissue, the central part of the plug, which still consists of non-flattened amoebocytes, becomes smaller. In 14-day-old wounds only a small core of amoebocytes is found, which is surrounded by a large mass of collagenous connective tissue in which amoebocytes and fibroblasts are present. Thirty days after the incision the cell plug has disappeared and only small groups of amoebocytes, intermingled with fibroblasts and infiltrated muscle cells, are left in the wound area. A number of the amoebocytes are loaded with residual bodies due to their high phagocytic activity. At 60–90 days, the architecture of the sub-epidermal connective tissue of the wound area is still different from that of the surrounding uninjured part of the body wall: a typical scar tissue is observed with masses of collagen in which blood as well as connective tissue cells are embedded (Fig. 25).

The first indications of the repair of the epidermis, which consists of one cell layer, are seen about 24 h after incision. Squamous cells grow out from the cut margins of the columnar epidermal epithelium. About 14 days later,

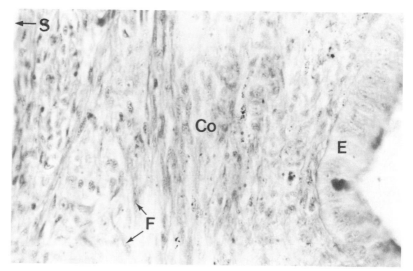

Fig. 25. Light micrograph of the wound area in *L. stagnalis* 60 days after incision, showing that the amoebocyte plug is replaced by scar tissue (collagenous connective tissue, Co). E, epidermis epithelium; F, fibroblasts; S, cephalopedal blood sinus. Azan stain. × 750.

the epithelial cells covering the plugs are again columnar in shape. At 30 days, a distinct basement membrane is observed under the restored epithelium. Furthermore, argyrophil connective tissue fibrils, obviously young collagen, are present in the area directly under the epithelium. This area is extended in older wounds.

These observations show that in gastropods, wound healing is exclusively a cellular process. Amoebocytes clear the wound area of cell debris and invading (micro)organisms, they clump together and close the wound, and they transform into fibroblasts which form a scar tissue of collagenous connective tissue.

It seems unlikely that plug amoebocytes form a special population of blood cells whose only function is the closure of wounds, since it has been established that (1) plug forming amoebocytes are morphologically similar to haemolymph and connective tissue amoebocytes, and (2) haemolymph amoebocytes which have previously phagocytosed foreign particles still take part in plug formation (Fig. 26; Sminia *et al.*, 1973).

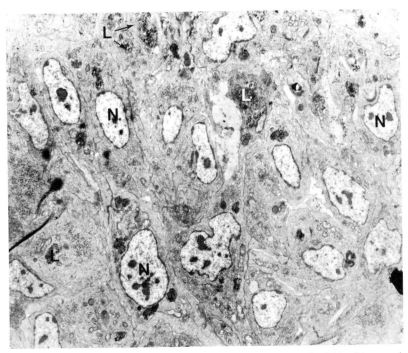

Fig. 26. Electron micrograph of part of an amoebocyte plug of a 24 h old wound of an ink-injected snail (*L. stagnalis*). The presence of ink-containing lysosomes (L) in plug amoebocytes indicates that phagocytosing amoebocytes also play a part in wound healing. N, nucleus. × 3000.

D. Other functions

Additional functions attributed to gastropod blood cells are involvement in digestion, distribution of food material to the organs, shell repair and regeneration (e.g. Wagge, 1951, 1955; Kapur and Gupta, 1970; Stang-Voss, 1970; Aboliņš-Krogis, 1973, 1976).

The assumed role of blood cells in digestion and transport of food material is mainly based on the occasional presence of fat droplets and glycogen particles within these cells. As, however (1) it has been shown that digestion of food is carried out by the digestive tract, in particular, intracellularly in the digestive gland (e.g. Meuleman, 1972; Walker, 1972), and (2) distribution and transport of food substances takes place via the plasma (e.g. Veldhuijzen and van Beek, 1976), the involvement of blood cells in these processes is highly unlikely. Probably, the presence of fat and glycogen reflects the active engagement of blood cells in defence and wound healing processes and are signs of the physiological state of the cells.

Blood cells are also said to play a role in shell repair and regeneration by transporting calcium and other substances from the digestive gland to the shell (e.g. Wagge, 1955; Kapur and Gupta, 1970; Aboliņš-Krogis, 1973, 1976). This hypothesis is chiefly based on the facts that (1) calcium-rich granules have been observed within blood cells and in lime cells, a particular cell type occurring in the digestive gland of terrestrial pulmonates (e.g. Walker, 1970) and (2) large numbers of blood cells occur at sites of shell repair (e.g. Aboliņš-Krogis, 1973, 1976). It seems, however, that the calcium-rich cells have been mistaken for blood cells; in fact they represent a special type of connective tissue cell, the calcium cell (e.g. Richardot and Wautier, 1972; Sminia et al., 1977). It thus seems unlikely that blood cells are involved in shell repair.

In conclusion, it is clear that the main functions of blood cells are cellular defence (phagocytosis and encapsulation) and tissue repair.

VI. Summary and concluding remarks

In the present chapter, several morphological and functional aspects of the circulatory system and of the blood cells of gastropods are reviewed and discussed.

The gross-morphology of the circulatory system, which is well documented in a number of gastropods, is described by presenting schematic drawings of the arterial and venous system of the freshwater snail, L.

stagnalis. Detailed information on the morphology of the walls of the blood vessels and sinuses appears to be lacking.

The morphology of the blood cells is described. There is controversy over the question of whether one or two blood cell types exist. Are the two blood cell types distinguished, expressions of one basic cell type or are they fully differentiated separate cell lines? Detailed investigations on the functions and differentiation of blood cells are required. However, such studies can only be fruitful when they are based on a sound knowledge of the morphology of the cells. It is proposed to classify the blood cells as spreading amoebocytes (granulocytes) and round cells (round amoebocytes and hyalinocytes).

Although blood cell proliferation has only been studied in detail in two snail species, it seems very likely that gastropods lack special, well-defined haemocytopoietic organs. Mature blood cells have the capacity to divide. Information on the development and differentiation of blood cells is lacking.

The main functions of blood cells are cellular defence (phagocytosis, encapsulation) and tissue repair. The role of the cells in these processes is described and discussed. Although there are indications that each blood cell has the capacity to perform diverse functions, there are differences: some cells are highly phagocytic, while others show only a slight phagocytic activity. Probably the morphological heterogeneity among blood cells reflects these differences in functional activities. It seems worthwhile to study the functional capabilities of blood cells in more detail.

It is suggested that two types of receptors are present on the plasma membranes of the phagocytic blood cells, viz., foreign body receptors and serum factor (opsonin, agglutinin) receptors. These receptors enable the blood cells to recognize and ingest foreign particles. The ingested material is digested in the lysosomal system in which large amounts of peroxidases are present. There are indications that the processing of foreign material by the phagocytes induces a non-specific augmentation of the cellular defence reactions. However, comprehensive studies, including quantitative approaches, of defence reactions are needed before conclusions can be drawn as to whether or not gastropods are able to acquire resistance.

Acknowledgements

The author wishes to thank Dr H. H. Boer for his valuable criticism during the preparation of the manuscript, Prof. Dr J. Lever, Dr Elisabeth Meuleman and Dr W. P. W. van der Knaap for reading the manuscript and Dr Benita Plesch for correcting the English text.

References

Abolinš-Krogis, A. (1973). *Z. Zellforsch. mikrosk. Anat.* **142**, 205–221.
Abolinš-Krogis, A. (1976). *Cell Tiss. Res.* **172**, 455–476.
Anderson, R. S. (1977). *In* "Comparative Pathobiology. Vol. 3. Invertebrate Immune Responses" (L. A. Bulla, Jr. and T. C. Cheng, Eds), pp. 1–20. Plenum Press, New York and London.
Anderson, R. S. and Good, R. A. (1976). *J. Invert. Pathol.* 27, 57–64.
Armstrong, D. E., Armstrong, J. L., Krassner, S. M. and Pauley, G. B. (1971). *J. Invert. Pathol.* **17**, 216–227.
Baecker, R. (1932). *Ergebn. Anat.* **29**, 449–585.
Bang, F. B. (1975). *In* "Invertebrate Immunity" (K. Maramorosh and R. E. Shopeof, Eds), pp. 137–151. Academic Press, New York and London.
Bayne, C. J. (1973). *J. comp. Physiol.* **86**, 17–25.
Bayne, C. J. (1974). *In* "Contemp. Topics in Immunobiology" (E. L. Cooper, Ed.), Vol. 4, pp. 37–45. Plenum Press, New York.
Bayne, C. J. and Kime, J. B. (1970). *Malacol. Rev.* **3**, 103–113.
Bekius, R. (1972). *Neth. J. Zool.* **22**, 1–58.
Boer, H. H. and Sminia, T. (1976). *Cell Tiss. Res.* **170**, 221–229.
Brooks, W. M. (1969). *In* "Immunity to Parasitic Animals" (G. J. Jackson, R. Herman and I. Singer, Eds), Vol. 1, pp. 149–171. Appleton-Century-Crofts, New York.
Brown, A. C. (1967). *Nature (Lond.)* **213**, 1145–1155.
Brown, A. C. and Brown, R. J. (1965). *J. exp. Biol.* **42**, 509–519.
Buchholz, K., Kuhlmann, D. and Nolte, A. (1971). *Z. Zellforsch. mikrosk. Anat.* **113**, 203–215.
Carriker, M. R. and Bilstad, N. M. (1946). *Trans. Amer. Microsc. Soc.* **65**, 250–275.
Carter, O. S. and Bogitsh, B. J. (1975). *Ann. N.Y. Acad. Sci.* **266**, 380–393.
Cheney, D. P. (1971). *Biol. Bull., Woods Hole* **140**, 353–368.
Cheng, T. C. (1975). *Ann. N.Y. Acad. Sci* **266**, 343–379.
Cheng, T. C. and Auld, K. R. (1977). *J. Invert. Pathol.* **30**, 119–122.
Cheng, T. C. and Galloway, P. C. (1970). *J. Invert. Pathol.* **15**, 177–192.
Cheng, T. C. and Garrabrant, T. A. (1977). *Int. J. Parasitol.* **7**, 467–472.
Cheng, T. C. and Rifkin, E. (1970). *In* "A Symposium on Diseases of Fishes and Shellfishes". *Amer. Fish. Soc. Spec. Publ.* **5**, 443–496.
Cheng, T. C., Thakur, A. S. and Rifkin, E. (1969). *Proc. Symp. Mollusca* **II**, 546–563.
Cheng, T. C., Chorney, M. J. and Yoshino, T. P. (1977). *J. Invert. Pathol.* **29**, 170–174.
Cheng, T. C., Guida, V. G. and Gerhart, P. L. (1978). *J. Invert. Pathol.* **32**, 297–302.
Cooper, E. L. (1976). "Comparative Immunology". Prentice-Hall Inc. Englewood Cliffs, New Jersey.
Cuénot, L. (1892). *Arch. Biol. (Liège)* **12**, 683–740.
Davies, P. S. and Partridge, T. (1972). *J. Cell Sci.* **11**, 757–769.
Faulk, W. P., Lim, H. K., Jeong, K. H., Heyneman, D. and Price, D. (1973). *In* "Non-Specific Factors Influencing Host Resistance" (W. Braun and J. Ungar, Eds), pp. 24–32. Karger Publishers, Basel, Switzerland.
Faust, E. C. (1920). *Bull. Johns Hopkins Hosp.* **31**, 79–84.
Fretter, V. and Graham, A. (1965). "British Prosobranch Molluscs". Adlard and Son Ltd., London.
Gebhart-Dunkel, E. (1953). *Zool. Jb. Alg. Zool. Physiol.* **64**, 235–266.

George, W. C. and Ferguson, J. H. (1950). *J. Morph.* **86**, 315–327..

Harris, K. R. (1975). *Ann. N.Y. Acad. Sci.* **266**, 446–464.

Harris, K. R. and Cheng, T. C. (1975a). *J. Invert. Pathol.* **26**, 367–374.

Harris, K. R. and Cheng, T. C. (1975b). *Int. J. Parasitol.* **5**, 521–528.

Haughton, I. (1934). *Quart. Jl. Microsc. Sci.* **77**, 157–166.

Hill, R. B. and Welsh, J. H. (1966). In "Physiology of Mollusca" (K. M. Wilbur and C. M. Yonge, Eds), pp. 125–174. Academic Press, New York and London.

Jeong, K. H. and Heyneman, D. (1976). *J. Invert. Pathol.* **28**, 357–362.

Kapur, S. P. and Gupta, A. S. (1970). *Biol. Bull., Woods Hole* **139**, 502–509.

Kinoti, G. K. (1971). *Parasitology* **62**, 161–170.

Kollmann, N. (1908). *Annls Sci. Nat. Zool. IX* **8**, 1–240.

Kostal, Z. (1969). *Biológia Bratisl.* **24**, 384–392.

Kress, A. (1968). *Revue suisse Zool.* **75**, 235–303.

Krupa, P. L., Lewis, L. M. and Vecchio, P. D. (1977). *J. Invert. Pathol.* **30**, 35–45.

Laryea, A. A. (1970). Studies on the nervous system of the slug *Agriolimax reticulatus*. Ph.D. Thesis, University of Wales.

Lever, J. and Bekius, R. (1965). *Experientia* **21**, 1–4.

Lie, K. J. and Heyneman, D. (1976). *J. Parasit.* **62**, 292–297.

Lie, K. J., Heyneman, D. and You, P. (1975). *J. Parasit.* **61**, 574–576.

Meuleman, E. A. (1972). *Neth. J. Zool.* **22**, 355–427.

Michelson, E. H. (1975). *In* "Invertebrate Immunity" (K. Maramorosch and R. E. Shope, Eds), pp. 181–195. Academic Press, New York and London.

Müller, G. (1956). *Z. Zellforsch. mikrosk. Anat.* **44**, 519–556.

Nold, R. (1924). *Z. wiss. Zool.* **123**, 373–430.

Pan, C. T. (1958). *Bull. Mus. comp. Zool. Harv.* **119**, 237–299.

Pan, C. T. (1965). *Ann. J. trop. Med. Hyg.* **14**, 931–976.

Pan, C. T. (1971). *Trans. Amer. Microsc. Soc.* **90**, 434–440.

Pauley, G. B., Granger, G. A. and Krassner, S. M. (1971). *J. Invert. Pathol.* **18**, 207–218.

Pemberton, R. T. (1974). *Ann. N.Y. Acad. Sci.* **234**, 95–121.

Plesch, B., Janse, C. and Boer, H. H. (1975). *Neth. J. Zool.* **25**, 332–352.

Prowse, R. H. and Tait, N. N. (1969). *Immunology* **17**, 437–443.

Reade, P. C. (1968). *Aust. J. exp. Biol. med. Sci.* **46**, 219–229.

Renwrantz, L. R. and Cheng, T. C. (1977a). *J. Invert. Pathol.* **29**, 88–96.

Renwrantz, L. R. and Cheng, T. C. (1977b). *J. Invert. Pathol.* **29**, 97–100.

Richardot, M. and Wautier, J. (1972). *Z. Zellforsch. mikrosk. Anat.* **134**, 227–243.

Schmid, L. S. (1975). *J. Invert. Pathol.* **25**, 125–131.

Schmidt, G. (1916). *Z. wiss. Zool.* **115**, 201–261.

Sminia, T. (1972). *Z. Zellforsch. mikrosk. Anat.* **130**, 497–526.

Sminia, T. (1974). *Cell Tiss. Res.* **150**, 443–454.

Sminia, T. (1977a). *Malacologia* **16**, 255–256.

Sminia, T. (1977b). *In* "Structure and Function of Haemocyanin" (J. V. Bannister, Ed.), pp. 281–288, Springer, Berlin, Heidelberg and New York.

Sminia, T. and Boer, H. H. (1973). *Z. Zellforsch. mikrosk. Anat.* **145**, 443–445.

Sminia, T. and Vlugt-van Daalen, J. E. (1977). *Cell Tiss. Res.* **183**, 299–301.

Sminia, T., Boer, H. H. and Niemandsverdriet, A. (1972). *Z. Zellforsch. mikrosk. Anat.* **135**, 563–568.

Sminia, T., Pietersma, K. and Scheerboom, J. E. M. (1973). *Z. Zellforsch. mikrosk. Anat.* **141**, 561–573.

Sminia, T., Borghart-Reinders, E. and van de Linde, A. W. (1974). *Cell Tiss. Res.* **153**, 307–326.

Sminia, T., de With, N. D., Bos, J. L., van Nieuwmegen, M. E., Witter, M. P. and Wondergem, J. (1977). *Neth. J. Zool.* **27**, 195–208.

Sminia, T., van der Knaap, W. P. W. and Edelenbosch, P. (1979a). *Devl. Comp. Immunol.* **3**, 37–44.

Sminia, T., van der Knaap, W. P. W. and Kroese, F. G. M. (1979b). *Cell Tiss. Res.* **196**, 545–548.

Stang-Voss, C. (1970). *Z. Zellforsch. mikrosk. Anat.* **107**, 141–156.

Stumpf, J. L. and Gilbertson, D. E. (1978). *J. Invert. Pathol.* **32**, 177–181.

Sudds, R. H. (1960). *J. Elisha Mitchell scient. Soc.* **76**, 121–133.

Tripp, M. R. (1961). *J. Parasit.* **47**, 745–751.

Tripp, M. R. (1970). *J. Reticuloendothel. Soc.* **7**, 173–182.

Veldhuijzen, J. P. and van Beek, G. (1976). *Neth. J. Zool.* **26**, 106–118.

Wagge, L. E. (1951). *Quart. Jl. microsc. Sci.* **92**, 307–321.

Wagge, L. E. (1955). *Int. Rev. Cytol.* **4**, 31–78.

Walker, G. (1970). *Protoplasma* **71**, 91–109.

Walker, G. (1972). *Proc. malac. Soc. Lond.* **40**, 33–43.

Wesenberg-Lund, C. (1939). "Biologie der Süsswassertiere. Wirbellose Tiere". Springer Verlag, Wien.

Wolburg-Buchholz, K. and Nolte, A. (1973). *Z. Zellforsch. mikrosk. Anat.* **137**, 281–292.

Yoshino, T. P. (1976). *J. Morph.* **150**, 485–494.

8. Bivalves*

T. C. CHENG

Marine Biomedical Research Program and Department of Anatomy (Cell Biology), Medical University of South Carolina, Charleston, South Carolina 29412, U.S.A.

CONTENTS

*The original information included in this contribution had resulted from research supported by a grant (PCM77-23785) from the National Science Foundation.

233

I. Introduction

Interest in the structure and functions of haemocytes of bivalve molluscs stems from two rationales: the acquisition of basic information, and gaining knowledge relative to the physiologic roles of these cells in economically important species, such as mussels, clams and oysters, which may benefit shell-fisheries. As a result of the contributions of numerous investigators, it is now known that these cells are involved in such important homeostatic functions as wound repair (Pauley and Sparks, 1965; des Voignes and Sparks, 1968; Pauley and Heaton, 1969; Ruddell, 1971a), shell repair (Wagge, 1951, 1955), nutrient digestion and transport (Yonge, 1923, 1926; Takatsuki, 1934a; Yonge and Nicholas, 1940; Zacks and Welsh, 1953; Wagge, 1955; Zacks, 1955; Owen, 1966; Purchon, 1968; Cheng and Cali, 1974; Cheng and Rudo, 1976a; Cheng, 1977), and excretion (Durham, 1892; Orton, 1923; Canegallo, 1924). In recent years, however, the major emphasis taken by investigators has been to understand the role of bivalve haemocytes in internal defence, i.e. what is commonly designated as cellular immunity.

Because of the vast amount of new information that has accumulated on the forms and functions of bivalve haemocytes, I agreed to write this review although it is recognized that earlier reviews are available (Andrew, 1965; Hill and Welsh, 1966; Cheng et al., 1969; Cheng and Rifkin, 1970; Narain, 1973).

All bivalves possess an open circulatory system which includes a heart. Furthermore, all possess leucocytes and some have haemoglobin-containing erythrocytes (Griesbach, 1891; Dawson, 1932; Sato, 1931; Hill and Welsh, 1966).

II. Structure of the circulatory system

The open circulatory system found in bivalve molluscs involves the movement of haemolymph not only in the heart and vessels but also seepage through tissues. In other words, the body tissues are continuously bathed in haemolymph. Haemolymph circulation is dependent, at least in part, on the heart, which receives haemolymph from venous sinuses and, as a result of contraction of its walls, pumps the haemolymph out of the heart through arteries which eventually empty into sinuses (Fig. 1). Excellent reviews of the physiology of the molluscan circulatory system have been contributed by von Brücke (1925) and von Buddenbrock (1965). Grassé (1960) has presented a review of the anatomy of circulatory systems as found in various groups of molluscs.

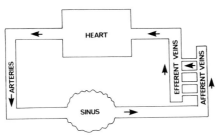

Fig. 1. Schematic diagramme showing general haemolymph flow pattern in a bivalve mollusc.

A. Heart

Typically, bivalves possess a three-chambered heart (Fig. 2), with an auricle on each side of the ventricle. The auricles are connected to the lateral surfaces of the ventricle by auriculoventricular apertures. In some species, such as the American oyster, *Crassostrea virginica*, the two auricles are superficially connected. The ventricle and auricles are enclosed within the pericardium, which is thin-walled. The rectum commonly runs through the ventricular chamber, although in oysters this posteriormost segment of the alimentary tract lies behind the heart. According to Franc (1960), the heart of members of the Aracea is duplicated. This is believed to be due to the large sizes of the byssus and posterior retractor muscle, i.e. two hearts are necessary to accommodate the massive amount of tissues.

The heart of bivalves is covered by an epithelial epicardium (Motley, 1933; Esser, 1934). Many glandular and darkly pigmented cells occur

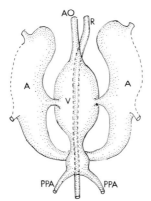

Fig. 2. Three-chambered heart of *Laevicardium crassum*. (Redrawn after White, 1942). A, auricle; AO, aorta; PPA, posterior pallial artery; R, rectum; V, ventricle.

between the epicardial cells covering the auricles. According to Franc (1960), this portion of the epicardium constitutes a part of the excretory system of bivalves. The pigmented cells, sometimes referred to as serous cells (or bodies) or brown cells, vary from light brown to almost black in colour. Collectively, these cells have been referred to as the "Pericardialdrüse" (pericardial glands) (Grobben, 1888), "das rotbraunes Organ" (Keber, 1851), or "Keber's organ" (Fernau, 1914). The possible function of these cells is considered later (pp. 278–279).

There is good evidence that the centres of automatation of the bivalve heart are diffuse. Based on the experiments of Jullien and Morin (1931) and Berthe and Petitfrère (1934a,b), it is concluded that the hearts of bivalves are myogenic, i.e. their intrinsic automatism originates in the muscular tissue and is not provoked by impulses from the central nervous system.

The normal movements of the bivalve heart serve as a pressure pump which must develop considerable power to propel the haemolymph through the circulatory system.

B. Haemolymph vessels

As stated earlier, bivalves have an open circulatory system. There are no capillaries joining the arterial and venous vessels; instead, haemolymph pumped from the heart into the arteries eventually seeps into sinuses. From these spaces the haemolymph enters veins and eventually is carried back to the heart (Fig. 1).

1. Arterial system

The complete circulatory system of only a few bivalves has been worked out. The major technical difficulty obstructing such anatomical studies is the fragility of the vessels. This condition renders the usual method of tracing vessels by injecting latex unusable. However, as a result of employing the techniques of Eble (1958), the main distribution pattern of the arterial system in *Crassostrea virginica* has been studied. This general pattern also probably holds true for other bivalves (Fig. 3).

Two large arteries emerge from the posteriodorsal aspect of the ventricle. The larger one is the anterior aorta, which upon leaving the heart, forms a bulbous enlargement. From this distended segment arises the large visceral artery. From the latter artery arise numerous branches, including the pericardial artery which supplies haemolymph to the pericardial wall.

The anterior aorta, as depicted in Fig. 3, gives rise to numerous primary and secondary branches among which the major ones are the gastric arteries which supply the digestive diverticula, the cephalic artery which supplies the anterior end of the body, and the labial artery which supplies the left and right labial palps.

The second aorta arising from the heart is the smaller posterior aorta (Fig. 3). It supplies haemolymph to the adductor muscle and rectum. The rectal artery arises from the posterior aorta at a point near the latter's origin. This artery follows the wall of the rectum.

2. Venous system

In *C. virginica*, the major haemolymph sinuses occur throughout the visceral mass, in the pallium, along the adductor muscle, and around the kidney. These are unlined spaces situated between adjacent tissues. Their shapes and dimensions vary, depending on the amount of haemolymph enclosed at the time.

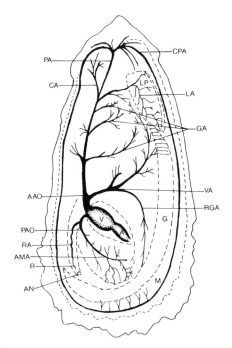

Fig. 3. Drawing of arterial system of *Crassostrea virginica*. AAO, anterior aorta; AMA, adductor muscle artery; AN, anus; CA, cephalic artery; CPA, circumpallial artery; G, gills; GA, gastric arteries; LA, labial palp artery; LP, labial palps; M, mantle; PA, common pallial artery; PAO, posterior aorta; R, rectum; RA, rectal artery; RGA, reno-gonadal artery; VA, visceral artery.

As depicted in Fig. 4, one of the major sinuses is the renal sinus. It is actually comprised of several smaller sinuses which surround the main part of the kidney. This sinus spreads into the adjacent connective tissues and is in communication with the internephridial passages leading to the pericardium.

Other major sinuses are the visceral sinus and the adductor muscle sinus (Fig. 4). The latter is situated below the renal sinus on the surface of the adductor muscle.

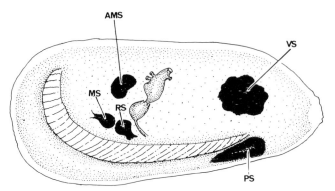

Fig. 4. Drawing showing locations of major haemolymph sinuses in *Crassostrea virginica*. AMS, adductor muscle sinus; MS, mantle sinus; PS, pallial sinus; RS, renal sinus; VS, visceral sinus.

There are two distinguishable venous systems: the afferent and efferent veins. The afferent system, which conducts deoxygenated haemolymph to the gills, consists of a single common afferent vein and two lateral afferent veins (Fig. 5). The common afferent vein runs along the ridge formed by the fusion of the two inner ascending gill lamellae. The haemolymph received by this vein comes from the deeper parts of the body and is brought by the (1) cephalic veins, (2) labial veins, (3) gastric and hepatic veins, (4) network of small reno-gonadal veins, (5) short renal vein and (6) adductor muscle vein (Fig. 5).

The paired lateral afferent veins are located along the axis of the outer ascending lamella where the lamella fuses with the mantle lobe (Fig. 5). These afferent veins receive haemolymph from the mantle through the pallial veins.

At regular intervals, the common afferent vein is connected with the lateral afferent veins by short transverse vessels (Fig. 5). The connection between the transverse vessels in gill tissues is maintained by vertical vessels which arise from the three afferent veins as a series in a double row, one

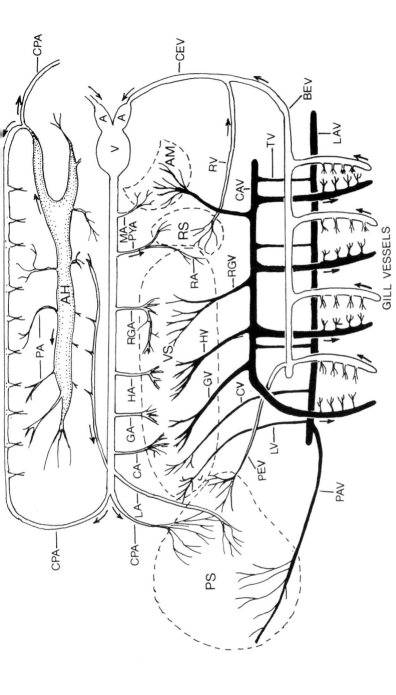

GILL VESSELS

Fig. 5. The circulatory system of *Crassostrea virginica* with only one demibranch and one accessory heart shown. A, auricle; AH, accessory heart; AM, adductor muscle; BEV, branchial efferent vein; CA, cephalic artery; CAV, common afferent vein; CEV, common efferent vein; CPA, circumpallial artery; CV, cephalic vein; GA, gastric artery; GV, gastric vein; HA, hepatic artery; HV, hepatic vein; LA, labial artery; LAV, lateral afferent vein; LV, labial vein; MA, adductor muscle artery; PA, pallial arteries; PAV, pallial afferent vein; PEV, pallial efferent vein; PS, pallial sinuses; PYA, pyloric artery; RA, renal artery; RGA, reno-gonadal artery; RGV, reno-gonadal vein; RS, renal sinuses; RV, renal vein; TV, transverse vein of gills; V, ventricle; VS, visceral sinuses.

following the inner, and the other the outer, lamella of the demibranch. At each interfilamentar shelf, the vertical vessels empty into a lacuna and eventually into the tubes of the ctenidial filaments. There is no special path for the return of haemolymph from the interfilamentar lamellae and the tubes as the filaments end as blind tubes.

The haemolymph channels in the interlamellar junctions communicate with the vertical vessels and it is through these channels that haemolymph flows from one lamella to the other. This flow is influenced by the contraction of the musculature of the gills and by contractions of the major afferent and efferent veins.

The efferent venous system (Fig. 5), which conducts oxygenated haemo-lymph to the heart, consists of two, short, common efferent veins which empty into the auricles, a pair of branchial efferent veins which run along the axis of the gill lamellae, the pallial efferent veins, and the interlamellar and interfilamental vessels of the gills.

The branchial efferent veins run along the gill axis parallel to the branchial nerves at the junctions of the ascending and descending lamellae. Along their course, these veins receive haemolymph from the renal sinuses and empty into the common efferent vein. Haemolymph circulating in the mantle is carried to the heart through the pallial sinuses and veins; however, some of the haemolymph from the posterior portion is drained back to the gills and to the branchial efferent vein.

C. Accessory hearts

Many species of bivalves have an accessory heart (Fig. 5). This organ is composed of a pair of pulsating vessels situated on the inner surface of the mantle fold. First described in the Pacific oyster, *Crassostrea gigas*, by Hopkins (1934, 1936), it serves to receive haemolymph from the nephridium and pump it to the gills and, against arterial pressure, into the marginal arteries of the mantle. This forces the mixed arterial and venous haemo-lymph into the small vessels of the mantle, which Hopkins considers to be a respiratory organ. Movement of haemolymph into the small mantle vessels is also governed by the rhythmic contractions of the mantle, which are also responsible for creating the respiratory currents in the mantle cavity (Redfield, 1917).

III. Origin and formation of haemocytes

A. Haemopoiesis

Surprisingly little is known about the origin of haemocytes in bivalves, or in all classes of molluscs for that matter. In the case of gastropods (see Chapter 7), Pan (1958) has suggested that in *Biomphalaria glabrata* these cells are formed from fibroblasts in the trabeculae of the haemolymph sinuses, similar cells in the wall of the saccular portion of the renal organ (which forms part of the pericardial sac), and in the loose connective tissue. More recently, Lie *et al.* (1975) have identified histologically the haemocyte-producing organ in this snail. It is situated between the pericardium and the posterior epithelium of the mantle cavity.

In the case of bivalves, Cuénot (1891) suggested that haemocytes may have their origin in special "glande lymphatiques" at the base of the gills. According to Narain (1973), K. Tanake in Japan considered the alimentary tract to be the centre of haemopoiesis in bivalves and suggested that the breakdown products of food, which he termed "monreal" or "life-substances," may be utilized for blood cell production by the cilia lining the intestinal wall. This unorthodox idea requires considerable supportive evidence before it can even be given any consideration.

Morton (1969) suggested that the circadian rhythm of adductor muscle activity and quiescence may be related to the formation of at least some of the haemocytes of *Dreissena polymorpha*.

Although no satisfactory account has yet been published as to where haemopoiesis occurs in bivalves, the generally accepted belief is that haemocytes arise from differentiation of connective tissue cells. This, however, is in need of verification and more sophisticated studies.

B. Ontogeny

Periodically some investigators interested in haemocytes of bivalves have expressed the opinion that there are different types of cells and at least some of these may represent ontogenetic stages of a same general type. Although I agree that there are different types of haemocytes, I am also in agreement with Feng *et al.* (1971) that whether different ontogenetic stages occur among circulating cells is still unknown.

Only two detailed accounts of the ontogeny of bivalve haemocytles have been published; that by Moore and Lowe (1977) and that by Mix (1976). In addition, Cheney (1969), in an unpublished dissertation, has presented some information. It is noted that all of these accounts are mostly based not on direct evidence but on interpretive evaluations.

Moore and Lowe (1977) recognized three categories of haemocytes in the mussel *Mytilus edulis*: (1) small basophilic hyaline cells or lymphocytes (Figs 6, 7), (2) larger basophilic haemocytes or macrophages (Figs 6, 7) and (3) eosinophilic granular haemocytes or granulocytes (Figs 6, 7). Among these,

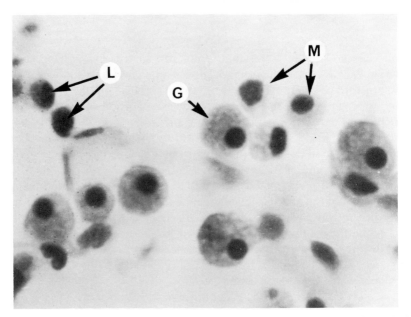

Fig. 6. Section of connective tissue of *Mytilus edulis* showing small basophilic hyaline cells (=lymphocytes) (L); large basophilic hyaline cells (=macrophages) (M); and granulocytes (G). (After Moore and Lowe, 1977, with permission of Academic Press, New York.) ×1700.

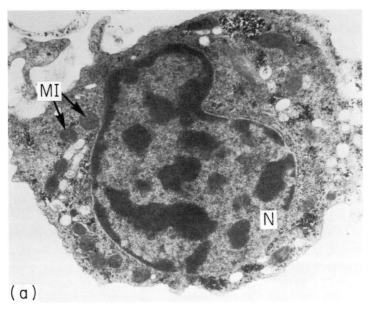

Fig. 7. Electron micrographs of haemocytes of *Mytilus edulis*. (a) A hyaline cell or lymphocyte showing nucleus (N) and mitochondria (MI). ×16 000.

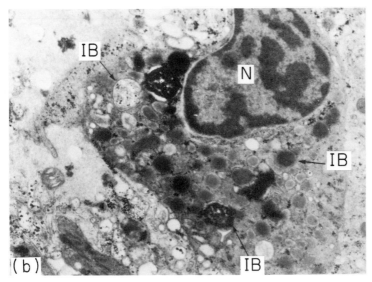

Fig. 7(b). A macrophage showing inclusion bodies (IB) of varying electron densities and nucleus (N) × 9000.

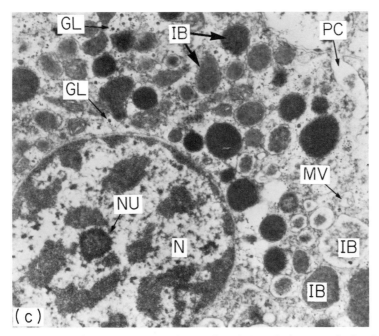

Fig. 7(c). Portion of granulocyte showing nucleus (N) enclosing a nucleolus (NU), membrane-bound inclusion bodies (IB), a pinocytotic channel (PC), a multivesicular body (MV), and glycogen granules (GL) in the cytoplasm. (All illustrations after Moore and Lowe, 1977, with permission of Academic Press.) × 14 000.

they have interpreted the occurrence of two developmental series: the basophilic haemocyte growth series and the acidophilic haemocyte series. It is noted that they studied cells in sections rather than smears.

The basophilic haemocyte growth series, according to Moore and Lowe (1977), contains a variety of structural types. Those cells, assumed to be the youngest are small (4–6 μm in diameter) and include a small volume of basophilic hyaline cytoplasm. Moreover, they are generally spherical and include a rounded nucleus. As the cytoplasm increases in volume, it becomes strongly basophilic and there is often an intense staining reaction with pyronin Y, indicating a high RNA content. Interpretatively, it was proposed that as these small basophilic hyaline cells (or lymphocytes) developed, they became macrophages (or larger basophilic haemocytes), which measure 7–10 μm in diameter. The RNA content in macrophages eventually decreases and the cells frequently take on an irregular appearance, produce pseudopods, and develop cytoplasmic granules and vacuoles. Also, the nuclei tend to lose their spherical shape and become irregular in outline.

The second growth series, the acidophilic line, is currently only represented by the eosinophilic granular haemocytes or granulocytes. According to Moore and Lowe (1977), although the majority of these cells measure 7–12 μm in diameter, a few smaller cells, each measuring 5–7 μm, are also observed. These smaller cells are believed to be young granulocytes and the degree of acidophilia of the cytoplasmic granules increases with growth. According to Moore and Lowe: "These cells (granulocytes) apparently constitute a distinct growth series, as no morphologic forms were observed which might have represented cells intermediate between macrophages and granular haemocytes. However, a few small granulocytes have been noted which may be intermediate between lymphocytes and granulocytes, although a number of granulocytes in the digestive gland connective tissue have been observed to have mitotic figures."

Mix (1976), who based his interpretations on the data of several investigators (Galtsoff, 1964; Tripp et al., 1966; Cheney, 1969; Cheng and Rifkin, 1970; Ruddell, 1971a,b,c; Feng et al., 1971; Foley and Cheng, 1972; Mix and Tomasovic, 1973) proposed that there are four "compartments" during what he termed "cell renewal" or, more appropriately, ontogenesis of bivalve haemocytes. As depicted in Fig. 8, following Cheney (1969), Mix postulated the occurrence of a leucoblast, which represents the "stem cell compartment." The leucoblast differentiates into a hyalinocyte, which represents the "proliferating compartment." Subsequently these hyalinocytes may differentiate into granular hyalinocytes or two classes of intermediates. These cells form what Mix has designated as the "maturing compartment" (Fig. 8). Further differentiation of the components of this compartment leads to the formation of basophilic granulocytes and acidophilic granulocytes from granular hyalinocytes, fibroblasts and myoblasts from one of the two intermediates, and pigment cells from the other intermediate. Finally, in what Mix considers "cell loss", all of the cells of the functional compartment contribute to defence reactions. Some are lost through senescence and diapedesis (Fig. 8).

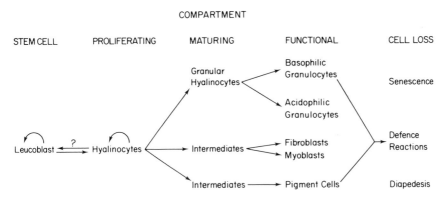

Fig. 8. Mix's (1976) generalized model of haemocyte ontogenesis in bivalve molluscs. Arrows indicate the flow of development from one compartment to another. (After Mix, 1976).

Mix's interpretations are not without merit, as a part of the experimental aspect of his report shows that stem-cell leucoblasts divide as do hyalinocytes of either the proliferating or the maturing compartment.

In the opinion of this reviewer, Mix's interpretation is too complex; furthermore, he has based his interpretation on the face values of others which are not always correct. A simpler and more realistic interpretation is presented later (see section on structure and classification of haemocytes—a synthesis).

IV. Structure and classification of haemocytes

Although the classification of bivalve haemocytes has been a topic of interest for many years, it has only been recently that some semblance of agreement has evolved; even then, complete agreement has still not been achieved. In Table I is tabulated the designations and characteristics of bivalve haemocytes that have been described.

TABLE I. Designations and characteristics of bivalve haemocytes.

Bivalve species	Haemocyte types (designations)	Dimensions (µm)	Diagnostic characteristics	Authority
Tapes decussata Mytilus edulis Ostrea edulis	coarse grained amoebocyte	8 × 12	birefringent granules; bluish colour/ colourless	Cuénot (1891)
Cardium norvegicum Cytherea chione Dreissena polymorpha	fine grained amoebocyte	8 × 12	birefringent granules; bluish colour/ colourless	
Unio requienii Unio sinuatus Anodonta cygnea	basophilic amoebocyte	8 × 12	small volume of cytoplasm; no granules	
Venus verrucosa Anomia ephippium Solen legumen Arca tetragona	erythrocyte	20 × 30	variable shape; reddish colour	
Mytilus edulis Ostrea edulis Unio pictorum	granulocyte		cytoplasmic granules neutrophilic tending to basophilic	De Bruyne (1895)
Anodonta cygnea	lymphocyte		small volume of agranular cytoplasm	
M. edulis Cardium edule Pecten maximus	granular cells: granulocytoblast granulocyte rhagioplast microlymphocyte		"spongy" cytoplasm granular cytoplasm metachromatic small amount of cytoplasm; essentially agranular	Betances (1921)
	plasma cell		moderate amount of finely granular cytoplasm; nucleus with spoke-like radiations	

Species	Cell type	Size (μm)	Description	Reference
M. edulis	eosinophilic granular haemocyte (designated as granulocyte)	7–12 diameter	include traces of certain lysosomal hydrolases and cytoplasmic granules (neutrophilic to acidophilic)	Moore and Lowe (1977)
	basophilic haemocyte lymphocyte (also designated as small basophilic hyaline cell)	4–6 diameter	small volume of basophilic hyaline cytoplasm; generally spherical	
	phagocytic macrophage (also designated as larger basophilic haemocyte)	7–10 diameter	include certain lysosomal hydrolases and irregular cytoplasmic granulation and vacuolation; many pseudopodia	
Mytilus coruscus	Type 1 amoebocyte	9–11 diameter	small basophilic nucleus 3–4 μm; cell round to square with moderate amount of granular or agranular cytoplasm	Feng et al. (1977)
	Type II amoebocyte		basophilic nucleus 4–5 μm; granular or agranular cytoplasm	
	Type III amoebocyte Type IV amoebocyte		round to oval basophilic nucleus 5–8 μm eosinophilic oval nucleus 5–10 μm; granular or agranular cytoplasm	
Crassostrea gigas Anadora subcrenata	I–R large	20–33·5 diameter	round or indefinite shape; eccentric nucleus, round, elliptical, or kidney-shaped; no nucleolus; abundant cytoplasm with no granules	Tanaka et al. (1961) and Tanaka and Takasugi (1964)
	I–R medium	8–16·2 diameter	round or indefinite shape; centric nucleus round or kidney-shaped; no nucleolus; scanty cytoplasm with no granules	
	I–R small	2–6·8 diameter	round cell; eccentric nucleus of indefinite shape; no nucleolus; scanty cytoplasm with no granules	
	I–S	4·5–9 diameter	indefinite shaped cell; eccentric nucleus of indefinite shape and without nucleolus; abundant cytoplasm with many coarse granules	

TABLE I.—*cont.*

Bivalve species	Haemocyte types (designations)	Dimensions (µm)	Diagnostic characteristics	Authority
	II large	60·5–64·8 diameter	centric nucleus round with one or no nucleolus; abundant cytoplasm light brown and with numerous round granules	
	II small	6·8–17·6 diameter	round, oval, elliptical, kidney-shaped, or indefinitely shaped cell; round eccentric nucleus with no nucleolus; moderate amount of cytoplasm with many granules	
	III large	59–70·2 diameter	round or kidney-shaped cell with round centric nucleus; nucleolus present; moderate amount of cytoplasm with many granules	
	III small	9·2–16·7 diameter	round, rectangular, or indefinitely shaped cell; eccentric or centric nucleus with no nucleolus; moderate amount of cytoplasm with many or few granules	
	V large	25·4–30·7 diameter	oval, round, or kidney-shaped cell; centric or eccentric nucleus with nucleolus; scanty cytoplasm with granules	
	V medium	16·8 diameter	kidney-shaped cell with eccentric nearly round nucleus without nucleolus; scanty cytoplasm with two granules	
	V small	5·3–9·3 diameter	oval or elliptical cell with eccentric round nucleus; nucleolus present; scanty cytoplasm with many granules	
	VI	4·8–69·4 diameter	round cell with centric nucleus without nucleolus; moderate amount of cytoplasm with no granules	

Species	Cell type	Size	Description	Reference
C. gigas	acidophilic granular amoebocyte		contains copper and a diazotized p-nitroaniline-positive material; with homogeneously electron-dense, membrane-lined cytoplasmic vesicles presence of electron-lucid vesicles with well-structured walls	Ruddell (1971a,c)
	basophilic granular amoebocyte agranular amoebocyte		torpedo-shaped (in wound plug): include clumped glycogen granules, lipid droplets and membranous whorls	Ruddell (1971b)
C. gigas	Type I amoebocytes		basophilic nucleus 3–5 μm surrounded by moderate amount of granular or agranular cytoplasm	Feng et al. (1977)
	Type II amoebocytes		neutrophilic nucleus 5–8 μm surrounded by granular or agranular cytoplasm	
	Type III amoebocytes		large eosinophilic nucleus 6–10 μm surrounded by large amount of granular or agranular cytoplasm	
Crassostrea virginica	amoebocyte, leucocyte or phagocyte			Stauber (1950)
C. virginica	granular cell or amoebocyte	6 diameter (unspread)	cytoplasmic granules neutrophilic or basophilic, bristle-like pseudopodia	Galtsoff (1964)
	hyaline cells	5–15 diameter	cytoplasm with few granules; slow moving; pseudopods lobose; basophilic	
C. virginica	granular leucocytes	10–20 diameter	nucleus round to oval; cytoplasm with light or dense granules	Feng et al. (1971)
	agranular leucocytes Type I	8 diameter	large oval, nucleus; scanty cytoplasm; little glycogen	
	Type II		oval nucleus; moderate amount of cytoplasm; clusters of glycogen granules	
	Type III		spherical nucleus with dense chromatin	

TABLE I.—cont.

Bivalve species	Haemocyte types (designations)	Dimensions (µm)	Diagnostic characteristics	Authority
C. virginica	granulocyte	$13 \pm 1 \times$ 12 ± 1.2 (fresh) $36 \pm 2 \times$ 23.5 ± 5 (spread)	with acidophilic, basophilic refractile granules, spike-like filopods	Foley and Cheng (1972)
	fibrocyte			
	primary fibrocyte	$9.3 \pm 1.6 \times$ 8.2 ± 1.5 (fresh)	with lobate nucleus and cytoplasmic vacuoles; few or no granules	
	secondary fibrocyte	$9.6 \pm 1.4 \times$ 9.4 ± 1.2 (spread)	with spherical or ovoid nucleus; few or no cytoplasmic granules but with vacuoles	
	hyalinocyte	9.3×8.2	scanty cytoplasm with none or few granules; pseudopodia lobose	
C. virginica	granulocyte	same[a]	same except that fibrocytes considered as terminal granulocytes	Cheng (1975)
	hyalinocyte	same[a]	same[a]	
C. virginica	granulocyte large (~10%)	13–20 diameter		Renwrantz et al. (1979)
	medium (40%)	9–13 diameter	same[a]	
	small (30%)	3–9 diameter		
	hyalinocyte	same[a]	same[a]	
Mercenaria mercenaria	granulocyte		with cytoplasmic granules and several hydrolases	Zacks (1955)

Species	Cell type	Size (μm)	Description	Reference
M. mercenaria	granulocyte	$35{\cdot}41 \pm 7{\cdot}66 \times 25{\cdot}71 \pm 6{\cdot}38$	granular cytoplasm includes vacuoles, acidophilic and basophilic granules and eccentric compact nucleus; forms filopodia	Foley and Cheng (1974)
	fibrocyte	$27{\cdot}43 \pm 7{\cdot}13 \times 20{\cdot}54 \pm 5{\cdot}78$	cytoplasm vacuolar; nucleus compact and eccentric; spike-like filopodia	
	hyalinocyte	$26{\cdot}18 \pm 6{\cdot}84 \times 21{\cdot}36 \pm 5{\cdot}19$	basophilic cytoplasm spread around large oval nucleus; few or no pseudopodia	
M. mercenaria	granulocyte	same[b]	cytoplasmic granules are true lysosomes; cells in terminal phase of physiologic cycle relative to degradation of phagocytosed materials include aggregates of glycogen and digestive lamellae in phagosomes; lipid-like droplets present	Cheng and Foley (1975) and Yoshino and Cheng (1976a)
	hyalinocyte	same[b]	rough endoplasmic reticulum present; absence of large numbers of electron-dense vesicles; electron-lucid membrane-bound vesicles present; some lipid droplets and glycogen granules present	
M. mercenaria	small granulocyte (61%)	$28 \pm 6{\cdot}49$ longest axis	cytoplasm with numerous granules of four types, some are lysosomes; highly motile, produces filopodia	Moore and Eble (1977)
	large granulocyte (37%)	$45 \pm 7{\cdot}3$ longest axis	cytoplasmic granules of four types, less abundant than in small granulocyte, some are lysosomes; produces filopodia	
	agranulocyte (2%)	$5 \pm 0{\cdot}68$ diameter	scant cytoplasm essentially devoid of granules; no motility	

TABLE I.—cont.

Bivalve species	Haemocyte types (designations)	Dimensions (µm)	Diagnostic characteristics	Authority
Ostrea edulis	granular leucocyte	9–13 diameter	nucleus compact, 3–4 µm diameter; cytoplasmic granules neutrophilic tending to basophilic; bristle-like pseudopodia	Takatsuki (1934a)
	lymphocyte	5–15 diameter	hyaline cytoplasm, no bristle-like pseudopodia; may include few granules	
O. edulis *Ostrea circumpicta* *Libitina japonica*	granulocyte		cytoplasmic granules neutrophilic tending to basophilic	Ohuye (1938a)
	lymphocyte		small volume of agranular cytoplasm	Ohuye (1938a,b)
Arca inflata	granulocyte		cytoplasmic granules neutrophilic tending to basophilic	Ohuye (1937)
	large lymphocyte	8–12 diameter	small volume of cytoplasm; few cytoplasmic granules	
	small lymphocyte	4–6 diameter	small volume of cytoplasm; no cytoplasmic granules	
Glycmeris resitius	coarsely granular granulocyte		cytoplasmic granules coarse	Ohuye (1937, 1938a)
	finely granular granulocyte		cytoplasmic granules fine	
	lymphocyte		small volume of agranular cytoplasm	
Anomia liscki	granulocyte		cytoplasmic granules acidophilic	Ohuye (1938a,b)
	lymphocyte		small volume of agranular cytoplasm	

Species	Cell type	Size	Description	Reference
Tapes semidecussata	granular leucocyte (granulocyte)	3–16 diameter	oval and eccentric nucleus; numerous acidophilic granules; spike-like pseudopods	Cheney (1971)
	hyaline leucocyte (leucocyte)	4–12 diameter	round nucleus; lobose pseudopodia	
Spisula solidissima	leucocyte		finely granular material around nucleus; vacuoles	Martin (1970)
Cardium norvegicum	finely granular eosinophil (48%)		round or oval nucleus; small eosinophilic granules concentrated around nucleus	Drew (1910)
	coarsely granular eosinophil (44%)		round or oval nucleus; with large well-defined eosinophilic granules	
	small basophil (8%)		small volume of cytoplasm; spherical nucleus	
Unio sp. _Anodonta_ sp.	amoebocyte		acidophilic cytoplasmic granules	Knoll (1893)
Anodonta grandis _Amblema costata_ _Quadrula quadrula_	granular eosinophilic amoebocyte (85%)	8–20 diameter	eosinophilic cytoplasmic granules; eccentric oval nucleus	Dundee (1953)
Uniomerus tetralasmus _Tritogonia verrucosa_ _Ligumia subrostrata_ _Lampsilis fallaciosa_	granular basophilic amoebocyte (7%)	7–12 diameter	basophilic cytoplasmic granules with occasional eosinophilic granules	
Proptera alata _Lasmigona complanata_ _Carunculina parva_	macronucleocyte (8%)	10–20 diameter	large nucleus: essentially agranular acidophilic cytoplasm	
Tritogonia verrucosa _Lampsilis purpurata_ _Actinonaias carinata_ _Megalonaias gigantea_ _Fusconaia ebena_ _Lasmigona complanata_	amoebocyte	5–25 diameter	darkly stained eccentric nuclei; coarse eosinophilic cytoplasmic granules	Motley (1933)

TABLE I.—*cont.*

Bivalve species	Haemocyte types (designations)	Dimensions (μm)	Diagnostic characteristics	Authority
Pectunclus sp.	amoebocyte		neutrophilic cytoplasmic granules	Knoll (1893)
Solen sp.	amoebocyte		acidophilic cytoplasmic granules	Knoll (1893)
Anadara inflata	amoebocyte	8 diameter	finely granular cytoplasm	Sato (1931)
Lamellidens corrianus	acidophil (26·32%)		largest; coarsely granular acidophilic cytoplasm; moderate number of filopodia	Narain (1968, 1969, 1972a)
	large basophil (60·2%)		medium size; finely granular basophilic cytoplasm; large number of filopodia	
	small basophil (13·48%)		smallest; finely granular basophilic cytoplasm; few filopodia	
Elliptio complanatum	acidophil (42·04%)	12–31·2 length	many acidophilic cytoplasmic granules	Hazleton and Isenberg (1977)
	large basophil (52·44%)	12–24 length	moderate number of basophilic cytoplasmic granules	
	small basophil (6·04%)	7·2–12 length	few basophilic cytoplasmic granules	

[a] See above Foley and Cheng (1972).
[b] See above Foley and Cheng (1974).

A. Existing classification schemes

Among the more detailed studies of bivalve haemocytes are those by Tanaka *et al.* (1961) on *Crassostrea gigas* and Tanaka and Takasugi (1964) on *Anadara subcrenata*. These Japanese investigators, taking into consideration the shapes and sizes of cells and nuclei, position and staining intensity of nuclei, distribution of chromatin, presence or absence of nucleoli, nature of perinuclear zone, and nature of the cytoplasmic granules, distinguished as many as six major types of cells (Table I). It is noted, however, that Tanaka *et al.* (1961) and Tanaka and Takasugi (1964) agree that all of the subcategories of cells fall under two broad types: granulocytic and agranulocytic.

More recently, Feng *et al.* (1971) contributed a light and electron microscope study of bivalve haemocytes. They classified cells of *C. virginica* as agranular and granular leucocytes, which in the presently accepted terminology, should be designated as hyalinocytes and granulocytes. In addition, Feng *et al.* recognized three categories of hyalinocytes and one type of granulocyte (Table I). The Type I hyalinocyte is a lymphocyte-like cell which is characterized by its relatively large and oval-shaped nucleus and scanty cytoplasm. Type II hyalinocytes possess an oval nucleus similar to that of Type I cells but have considerably more cytoplasm. Type III hyalinocytes are characterized by a spherical nucleus containing dense chromatin.

Although Feng *et al.* (1971) recognized one type of granulocyte, they believed that there are three distinct types of cytoplasmic granules which when stained with Giemsa's stain, appear refractile, dark blue and pink. These investigators have attempted to correlate their Types A, B and C granules as observed with electron microscopy with the three differently staining types. They proposed that Type A granules are refractile, Type B granules stain dark blue, and Type C granules stain pink in Giemsa-stained preparations. Granulocytes may include one or more types of granules.

Cheng and Cali (1974) and Cheng *et al.* (1974) also studied the granulocytes of *C. virginica* with the electron microscope. Unlike Feng *et al.* (1971), they interpreted the cytoplasmic granules of these cells to be only of one type. Incidentally, since these "granules", as will be reviewed in the next section, are known to be primary phagosomes, Cheng and Cali (1974), Cheng *et al.* (1974) and Cheng (1975) have referred to them as "vesicles".

Although it is now generally accepted that only two categories of haemocytes occur in *C. virginica*, recent studies by Renwrantz *et al.* (1979) revealed three groups of granulocytes which are distinguishable by their dimensions (Table I). It is still uncertain whether the small and medium cells are younger granulocytes, although it has been ascertained that the large and medium granulocytes are more phagocytic (Renwrantz et al., 1979).

Cheney (1971), in addition to summarizing what was known at that time about the morphology of invertebrate haemocytes, also contributed a study of the haemocytes of the Manila clam, *Tapes semidecussata*. He concluded that there are two general types: hyaline leucocytes (which he also termed leucocytes) and granular leucocytes (which he also termed granulocytes) (Table I). It is important to note that Cheney stated that: "Cells in these categories showed extreme variations in size and staining characteristics. A decrease in nuclear size often was correlated with an increase in nuclear heterochromasia and eccentricity and cytoplasmic granularity".

Foley and Cheng (1972) re-examined the cell types in the American oyster, *C. virginica*. On the basis of dimensions and behaviour of living cells, they concluded that there are two size classes of haemocytes: large and small cells. Furthermore, the large cells can be distinguished as granulocytes and fibrocytes. The small cells, which are either agranular or slightly granular, were designated as hyalinocytes. Thus, three haemocyte types were recognized (Table I). Subsequently, they reported similar cell types in *Mercenaria mercenaria* (see Foley and Cheng, 1974) (Table I).

The granulocytes of *M. mercenaria* are characterized by the presence of an abundance of cytoplasmic granules which are primarily limited to the endoplasm (Fig. 9). Foley and Cheng (1974) recognized four types of cytoplasmic inclusions: (1) vermiform bodies, which are thin and elongate, reaching about 2 μm in length; (2) vacuoles from less than 1 to 4 or 5 μm in diameter and which may contain inclusions; (3) small, spherical refractile inclusions each averaging 0·7 μm in diameter and which are most readily recognizable when observed with Nomarski interference optics; and (4) granules of variable shape, varying from spherical to elongate, and averaging 0·8 μm in greatest diameter (Fig. 9).

In granulocytes that have spread against a substrate (Fig. 9), the peripheral undulating ectoplasm is essentially agranular. They produce spike-like pseudopods each measuring 1–8 μm in length. Spread granulocytes are motile and portray active cyclosis.

What Foley and Cheng (1974) had designated as fibrocytes are spindle-shaped when contracted, and subsequent to adherence, spread slowly across the substrate, with the nucleus usually located along one side (Fig. 10). These cells contain motile vermiform bodies, each measuring up to 2 μm long, and which are more or less evenly distributed throughout the cytoplasm. Small refractile bodies (< 1 μm) generally occur adjacent to the nucleus.

Beyond the periphery of the spread fibrocyte protrude a variable number of filopodia, which like those of granulocytes, can be traced back into the

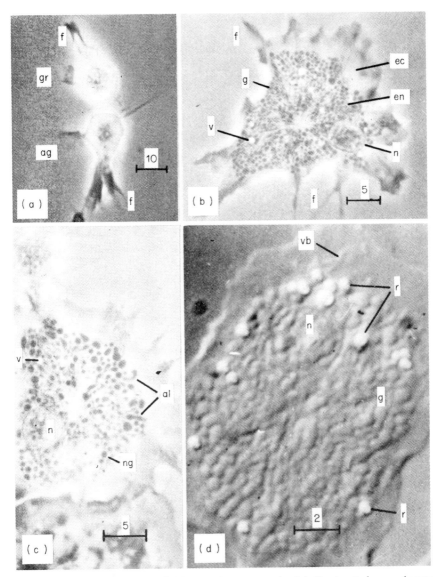

Fig. 9. Living granulocytes of *Mercenaria mercenaria*. (a) Contracted granulocyte (gr) and hyalinocyte (ag) with filopodia-like projections (f) (phase contrast). (b) Spread granulocyte showing separation of endoplasm (en) and ectoplasm (ec), filopodia (f), and various inclusions, including granules (g), vacuoles (v) and nucleus (n) (phase contrast). (c) Spread granulocyte enclosing both spherical (ng) and elongate (al) granules (phase contrast). (d) Spread granulocyte showing vermiform body (vb) in ectoplasm and nucleus (n), refractile bodies (r) and granules (g) in endoplasm (Nomarski). Scale-bar units in μm. (After Foley and Cheng, 1974.)

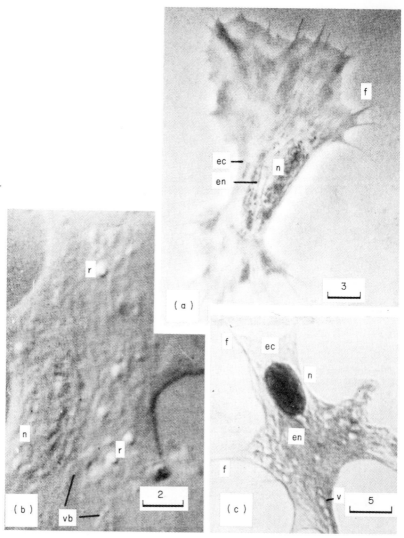

Fig. 10. "Fibrocytes" of *Mercenaria mercenaria*. (a) A living cell showing ectoplasm (ec) with filopodial projections (f), endoplasm (en) and nucleus (n) (phase contrast). (b) Living cell containing refractile bodies (r), nucleus (n) and vermiform bodies (vb). (Nomarski interference). (c) Stained cell showing filopodia projecting from ectoplasm (ec), and containing nucleus (n) and vacuoles (v) in endoplasm (en). (Giemsa stain, Nomarski interference). (All figures after Foley and Cheng, 1974; with permission of *Biological Bulletin*.) Scale-bars in μm.

endoplasm. Following spreading, small sections of cytoplasm are occasionally detached from the substrate and contract and undulate, resulting in the slow movement of the cell across the slide. Cyclosis also occurs in these cells.

Hyalinocytes contain some vermiform bodies, vacuoles, and few refractile bodies located near the large, compact, eccentrically situated nucleus (Fig. 11). Only a few or, as is generally the situation, no filopods are produced. Pseudopodia are lobose. Slow cyclosis does occur.

Subsequent to the earlier study involving light microscopy (Foley and Cheng, 1974), Cheng and Foley (1975) examined the haemocytes of *M. mercenaria* with transmission electron microscopy (Figs 12–14). Cells designated by Foley and Cheng (1974) as fibrocytes often included large aggregates of glycogen granules, mostly as α rosettes (Fig. 12). Furthermore, large, electron-lucid, membrane-delimited vesicles occur in the cytoplasm and these commonly enclose arrays of concentric lamellae (digestive lamellae) and partially degraded cellular constituents (Fig. 12). These larger, electron-lucid vesicles are phagosomes. The mitochondria of fibrocytes, like those of granulocytes (Fig. 13) and hyalinocytes (Fig. 14), have relatively few cristae. Also present are small, electron-lucid, membrane-bound vesicles

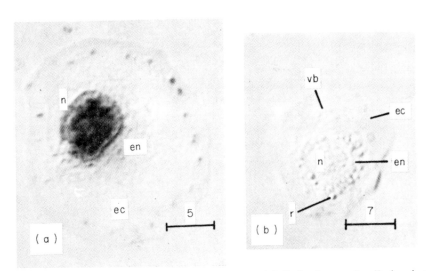

Fig. 11. Hyalinocytes of *Mercenaria mercenaria*. (a) Stained spread cell showing ectoplasm (ec) with no projecting filopodia and endoplasm (en) with nucleus (n). (Giemsa stain; Nomarski interference). (b) Living cell showing ectoplasm (ec), endoplasm (en), refractile bodies (r) and vermiform bodies (vb). (Nomarski interference). Bar-scale in μm. (Both figures after Foley and Cheng, 1974; with permission of *Biological Bulletin*.)

which may or may not include amorphous material. Those that enclose such material are most probably lysosomes.

As a result of finding large aggregates of glycogen granules and phagosomes that enclose digestive lamellae surrounding partially degraded cellular constituents (Fig. 12), Cheng and Foley (1975) proposed that fibrocytes are actually granulocytes that have undergone degranulation and are carrying out intracellular digestion. The basis for this interpretation is that granulocytes of bivalves, specifically those of the oyster, *C. virginica*, which have endocytosed bacteria, form digestive lamellae in the phagosomes around the microorganisms which eventually become enzymatically degraded (Cheng and Cali, 1974). It is also known that the intracellular digestion of bacteria, among other phenomena, results in the isolation of their carbohydrate constituents, which are converted to glycogen by some

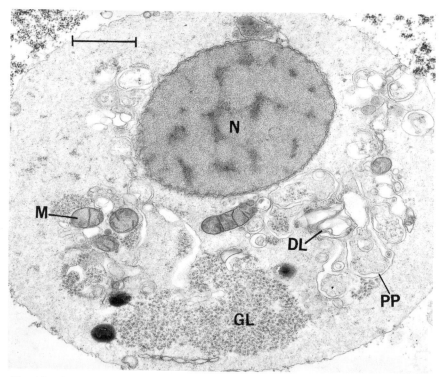

Fig. 12. Electron micrograph of *Mercenaria mercenaria* "fibrocyte". DL, digestive lamellae; GL, glycogen granules; PP, primary phagosome; M, mitochondrion; N, nucleus. (After Cheng and Foley, 1975; with permission of Academic Press.) Scale-bar = 2 μm.

yet undetermined pathway and this polysaccharide is eventually deposited in large aggregates within granulocytes (Cheng and Cali, 1974). This fact, in addition to others, was also employed as a basis for considering what had been interpreted to be fibrocytes to be granulocytes at a later stage of the endocytosis-intracellular degradation process since, as stated, electron microscopy has revealed large masses of glycogen granules in what Foley and Cheng (1974) had originally designated as "fibrocytes". Consequently, it now appears that there are two categories of haemocytes in both *M. mercenaria* and *C. virginica* (see Cheng, 1975), granulocytes and hyalinocytes. This indicates that the classification of molluscan haemocytes based purely on morphological differences as determined by light microscopy is unsatisfactory.

As reported earlier (see ontogeny), Moore and Lowe (1977), utilizing both light and electron microscopy, found two main cell lines in the mussel, *Mytilus edulis*: basophilic haemocytes and eosinophilic granular haemocytes (Table I, Figs 6, 7); the latter were also designated as granulocytes.

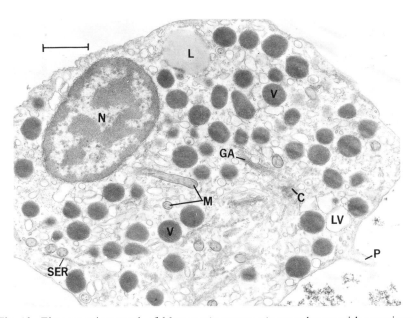

Fig. 13. Electron micrograph of *Mercenaria mercenaria* granulocyte with prominent centrosome. C; GA, Golgi apparatus; L, lipid droplet; LV, lucid vesicle; M, mitochondria; N, nucleus; P, pseudopodium; SER, smooth endoplasmic reticulum; V, electron-dense vesicle. (After Cheng and Foley, 1975; with permission of Academic Press.) Bar = 2 μm.

Furthermore, they subdivided the basophilic haemocytes into (1) small lymphocytes and (2) larger phagocytic macrophages. The latter were shown by cytochemistry to include certain lysosomal hydrolases. It is noted, however, that they also reported traces of acid phosphatase, β-glucuronidase, β-hexosaminidase and indoxyl esterase in granulocytes although not at the levels of those which occur in macrophages.

From the nomenclatural standpoint, Moore and Lowe (1977) employed "small basophilic hyaline cells" and "lymphocytes" interchangeably as was the case with the designations "larger basophilic haemocytes", "macrophages", "eosinophilic granular haemocytes" and "granulocytes".

Another recent study is that of Moore and Eble (1977) who also examined the haemocytes of the hard clam, *M. mercenaria*. By studying cells in smear preparations, they have concluded that there are three types: agranulocytes, small granulocytes and large granulocytes (Table I).

Feng *et al.* (1977) contributed a light and electron microscope study of the haemocytes of two marine bivalves from Japan, *Mytilus coruscus* and

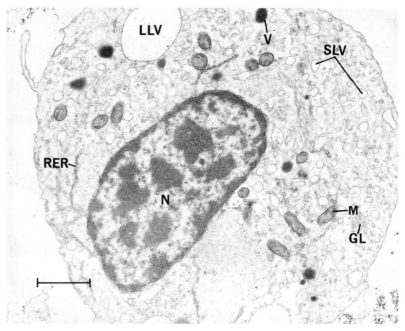

Fig. 14. Electron micrograph of *Mercenaria mercenaria* hyalinocyte. GL, glycogen granules; LLV, large lucid vesicle; M, mitochondrion; N, nucleus; RER, rough endoplasmic reticulum; SLV, small lucid vesicles; V, electron-dense vesicle. (After Cheng and Foley, 1975; with permission of Academic Press.) Bar = 2 μm.

Crassostrea gigas. Based on nuclear morphology, they concluded that there are four types of cells in *M. coruscus*, with each type including both agranular and granular cells (Table I; Fig. 15). In fact, they have designated those with cytoplasmic granules as granulocytes. This, unfortunately, has led to confusion relative to nomenclature.

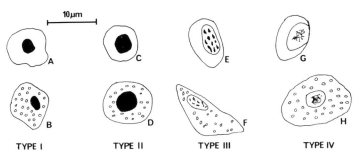

Fig. 15. Drawings of the four types of haemocytes of *Mytilus coruscus* as defined by Feng *et al.* (1977). Note that these authors consider each type to include both agranular and granular cells. (Redrawn after Feng *et al.*, 1977.)

As a result of their electron microscope studies, Feng *et al.* (1977) have identified five types of granules in *M. coruscus* granulocytes as well as lipid-like bodies. All of the granules are membrane-delimited.

Type A granules (Fig. 16), measuring 0·45–1·1 µm, include mostly fine particles of light density which are somewhat homogeneous. Type B granules (Fig. 16), each measuring 0·45–1·64 µm, are spherical and include a very dense round core embedded in a fairly dense matrix of finely granular material. This is the most common type of granule. Type C granules (Fig. 16) are typically irregular in their outlines and vary greatly in size from 0·72 to 2·29 µm. They also include a very dense core which is centrally or eccentrically located. The area between the core and the outer membrane, however, is largely translucent, with or without sparse to moderate amounts of coarsely granular material. Type D granules (Fig. 16) are characterized by their large size, 1·71–2·45 µm, and the more or less ellipsoidal shape of both their outline and dense core. Furthermore, the zone between the surface membrane and the core is more or less filled with coarse particles. Type E granules (Fig. 17) are the smallest of the cytoplasmic granules. Each measures 0·18–0·47 µm and is in the form of spherical dense bodies. They are the major type of granules in granular cells of *M. coruscus*.

Also basing their decision on nuclear morphology, Feng *et al.* (1977) reported the occurrence of three types of haemocytes in *C. gigas*. Again,

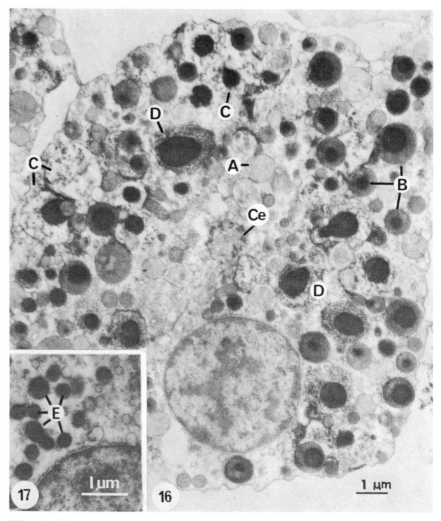

Fig. 16. Electron micrograph of a granular haemocyte (granulocyte) of *Mytilus coruscus* including A, B, C and D Type granules. A centriole (Ce) is also present. (After Feng *et al.*, 1977, with permission of Plenum.) × 13 000.

Fig. 17. Electron micrograph of a granulocyte of *Mytilus coruscus* including Type E granules (E). (After Feng *et al.*, 1977, with permission of Plenum.) × 19 000.

these investigators reported that each type includes both agranular and granular cells (Table I).

Their electron micrographs have revealed, among other features, that a nucleolus occurs in some cells, not all. Also, although there is usually only one kind of cytoplasmic granule, some cells include two of three different types which they designated Type A, Type B and Type C[1] granules.

Earlier, Ruddell (1971a) had also studied the fine structure of *C. gigas* haemocytes. He concluded that there are two types of granular amoebocytes: acidophils and basophils (Table I). Acidophils are characterized by the inclusion of relatively small, spherical, amorphous, osmiophilic, cytoplasmic granules, each measuring 0·4–0·56 μm in diameter. The matrix of each granule is packed with an amorphous, finely granular material. Basophils are characterized by the inclusion of large granules, each measuring 0·7–1·2 μm in diameter.

According to Feng *et al.* (1977), the Type A granules in *C. gigas* resemble Ruddell's (1971a) basophilic granules, and their C[1] granules resemble Ruddell's acidophilic granules.

Although there are similarities between these granules, this reviewer, as a result of his own electron microscope studies on the granules or vesicles of the related oyster *C. virginica* (see Cheng and Cali, 1974; Cheng *et al.*, 1974; Cheng, 1975), is of the opinion that the Type B granules of Feng *et al.* (1977) and the acidophilic granules of Ruddell (1971a) are most probably lysosomes. This interpretation is supported by the finding of Feng *et al.* (1977) that these granules are rich in acid phosphatase. Furthermore, this reviewer believes that Ruddell's (1971a) basophilic granules represent Cheng and Cali's (1974) secondary phagosomes and what have been described by Feng *et al.* (1974) as the cytoplasmic granules of granulocytes of *C. virginica*. This interpretation is based on similarities in sizes and the complex structure of the wall of each granule (or vesicle).

As the information pertaining to intracellular digestion in oyster granulocytes is reviewed in the subsequent section, these interpretations will become more understandable.

It is noted that Ruddell (1971b) also studied the fine structure of what he considered to be agranular amoebocytes of *C. gigas*. This study was not conducted on isolated cells but on histological sections of cells comprising plugs formed during wound healing. This reviewer believes that the cells depicted by Ruddell are not agranular cells but granulocytes which were at the terminal stages of intracellular degradation of endocytosed cellular debris. This opinion is based on the identification of large aggregates of glycogen granules in the cytoplasm of the cells (see Ruddell, 1971b, Fig. 2) as well as digestive lamellae (see Ruddell, 1971b, Fig. 6). These features, as stated earlier, are characteristic of "spent" granulocytes of oysters.

B. A synthesis

It is apparent from the above that there has been a variety of interpretations as to how many types of haemocytes occur in bivalve molluscs. Furthermore, there has been little agreement as to the designations of these cells. As a result of analysing those criteria that have been employed for distinguishing between cell types, it has become apparent to this reviewer that most are artifactual. The fault lies, not with the earlier investigators, but with unavailability of physiological information until only recently with the advent of electron microscopy.

When the classification systems for bivalve haemocytes employed by various authors up to the present are analysed, it becomes apparent that there are two principal patterns. The first, attributable to Takatsuki (1934a), involves classifying these cells into two categories: granular cells, i.e. those with conspicuously large numbers of cytoplasmic granules, and hyalinocytes, i.e. those with less cytoplasm and including none or only a few granules. The second system, originated by Cuénot (1891), recognizes three categories of cells: finely granular haemocytes, coarsely granular haemocytes and cells with very little cytoplasm surrounding the nucleus (Table I). All subsequent classification systems may be considered modifications of one of these two systems.

Having evaluated the descriptions of all of those who have basically adopted the Cuénot system, this reviewer rejects a system which involves differentiating between coarsely and finely granular cytoplasm. The formerly employed system of distinguishing between several types of cytoplasm, viz., homogeneous, granular, alveolar, fibrillar, and subcategories of these (Jordon, 1952), is no longer considered valid since the morphology of cytoplasm at the light microscope level is readily altered by employing different fixatives. Furthermore, electron microscope studies on several species of bivalves, namely *M. edulis* by Moore and Lowe (1977); *M. coruscus* by Feng *et al.* (1977); *C. virginica* by Rifkin *et al.* (1969), Cheng and Rifkin (1970) and Feng *et al.* (1971); *C. gigas* by Ruddell (1971a,b,c) and Feng *et al.* (1977); *M. mercenaria* by Cheng and Foley (1975) and Yoshino and Cheng (1976a) have revealed that the cytoplasmic "granules" are functional organelles such as mitochondria, lysosomes, phagosomes, Golgi, etc. Variations in numbers of one type of organelle over another, coupled with the method(s) of fixation, could result in morphological differences at the light microscope level. Also, differences in the dimensions of cytoplasmic granules could reflect ontogenesis or different stages during the physiologic cycle of a cell. For example, as reviewed earlier, what had been considered to be "fibrocytes" among the circulating haemocytes of *C. virginica* and *M. mercenaria* are now considered to be degranulated granulocytes at the terminal stages of the endocytosis-intracellular degradation process. If this was not known, a classification based on static morphology would no doubt consider "fibrocytes" as a distinct type of cell.

Cell dimensions have also been employed as a criterion for distinguishing between types of haemocytes; however, cell sizes by themselves are not a totally reliable criterion, since dimensions of haemocytes measured *in vitro* vary considerably depending upon the suspension medium (Narain, 1972a; Foley and Cheng, 1972, 1974; Cheng and Auld, 1977). Furthermore, cells designated as hyalinocytes, hyaline cells, or lymphocytes are not only smaller than granulocytes (Table I) but they also behave differently when placed on a solid substrate. Specifically, they do not produce semipermanent filopodia and when migrating, hyalinocytes produce small numbers of lobopodia. Hyalinocyte is the preferred designation for this type of cell since employment of the term "lymphocyte" infers that there is a lymphatic system in bivalves comparable to that of vertebrates when there is none. There is no doubt that what Dundee (1953) has described as macro-nucleocytes in a number of freshwater bivalves (Table I) are hyalinocytes.

It should also be emphasized that real size differences may occur among structurally similar cells. It has been pointed out by Renwrantz *et al.* (1979), using a technique which prevents spreading, that there are three size-classes of granulocytes in *C. virginica*; furthermore, they have determined that the large- and medium-sized granulocytes are more phagocytic. Rather than classify the three size-classes of granulocytes as being distinct, this reviewer considers them to be of one type but at different developmental stages. The same holds true for the so-called Type I, II and III granular cells of *Mytilus coruscus* as described by Feng *et al.* (1977); the I-R large, I-R medium, and I-R small and the II large and II small cells of *C. gigas* as described by Tanaka *et al.* (1961); the small and large granulocytes of *M. mercenaria* as described by Moore and Eble (1977); and the large and small lymphocytes (=hyalinocytes) of *Arca inflata* as described by Ohuye (1937) (Table I).

For the reasons stated, the classification system initiated by Cuénot (1891) based on finely or coarsely granular cytoplasm is believed to be artificial. On the other hand, this reviewer accepts what may be designated as Takatsuki's (1934a) system, i.e. consider the haemocytes of bivalves to be of two categories: granulocytes and hyalinocytes. However, this system has also been modified because of differences in the staining affinities of granulocytes. Specifically, some authors (Dundee, 1953; Narain, 1968, 1969, 1972a; Ruddell, 1971a,c; Hazleton and Isenberg, 1977) considered granulocytes with acidophilic granules to be distinct from those with basophilic, and sometimes neutrophilic granules. This is understandable if one classifies these cells based on static morphology. However, as a result of critical microscopy on a large number of identically treated cells, Galtsoff (1964),

Feng *et al.* (1971), and Foley and Cheng (1972) studying *C. virginica*; Ohuye (1938a) studying *Ostrea circumpicta*; Foley and Cheng (1974) with *M. mercenaria*; Ohuye (1937) studying *Arca inflata*; and Dundee (1953) with *Uniomerus tetralasmus*, *Tritogonia verrucosa* and *Ligumia subrostrara*, reported that granulocytes may include two or three combinations of acidophilic, basophilic and neutrophilic granules. Thus, it would appear that to draw sharp demarcations between types of granulocytes based on the staining affinities of the granules is artificial. In some species, many of these granules (vesicles) include lysosomal elements (Cheng and Cali, 1974; Cheng, 1975; Yoshino and Cheng, 1976a; Moore and Eble, 1977), some of which may have an alkaline milieu whilst others may be acidic for the functioning of alkaline and acid phosphatases, respectively. Variations in granule staining affinities may also reflect their ontogeny in which they may go through an alkaline phase prior to becoming acidic. This hypothesis is supported by the fact that granulocytes which include a mixture of basophilic and acidophilic granules usually include considerably more of one kind, which suggests that the transition from one phase to another is rapid. It is noted that usually it is cells with large numbers of acidophilic granules that include a few basophilic ones. I am of the opinion that this hypothesis is the more acceptable one; furthermore, I believe that young granulocytes include basophilic granules and as they mature, their granules become acidophilic.

It is of interest to note that Moore and Lowe (1977) have reported that both the small basophilic haemocytes (4–6 μm) and larger basophilic haemocytes (7–10 μm) of *M. edulis* are smaller than the eosinophilic granular haemocytes (7–12 μm). Although these investigators have interpreted the small basophilic haemocytes, which they also designated as lymphocytes, to be the precursors of the larger basophilic haemocytes, which they also termed phagocytic macrophages, this reviewer, having examined their descriptions and micrographs in detail, is of the opinion that the cells which they called lymphocytes (or small basophilic hyaline cells) are in fact hyalinocytes. What they designated as phagocytic macrophages (or larger basophilic haemocytes) are probably young granulocytes (or eosinophilic granular haemocytes). This interpretation is based on the following: (1) both the "phagocytic macrophages" and the "granulocytes" include certain lysosomal hydrolases, i.e. acid phosphatase, β-glucuronidase, β-hexosaminidase and indoxyl esterase, although the amounts in granulocytes are less. These lysosomal enzymes are characteristically present in granulocytes. (2) Electron micrographs by Moore and Lowe of "macrophages" (1977, Figs 9, 10) (Fig. 7) reveal the typical cytoplasmic granules, which are, in fact, lysosomes and phagosomes enclosing endocytosed foreign materials. These, as reviewed earlier, are diagnostic characteristics of granulocytes. (3) Their electron micrographs of granulocytes (Moore and Lowe, 1977, Figs 11, 12) (Fig. 7) are typical of the granulocytes of other bivalves, such as *M. mercenaria* (see Cheng and Foley, 1975) and *Mytilus coruscus* (see Feng *et al.*, 1977).

In view of the above, it may be concluded that Moore and Lowe (1977) have found what are apparently the two categories of haemocytes common to all bivalves; hyalinocytes and granulocytes. Their smaller "phagocytic macrophage," which has basophilic granules, are younger (smaller) granulocytes.

Other characteristics that have been employed to distinguish between types of haemocytes are the acidophilic or basophilic nature of the nucleus and cytoplasm (e.g. Dundee, 1953; Feng *et al.*, 1977). Again, it is questionable whether these can be employed as reliable criteria for separating types of cells, since they may alter with either the fixative used (Foley and Cheng, 1972) or perhaps also with the age of the cells.

It now appears that there is solid basis for the recognition of only two categories of cells: granulocytes and hyalinocytes. However, as depicted in Fig. 18, there are detectable staining and other morphological alterations which occur during ageing and functional shifts during the life of cells. The following are interpretive ontogenetic lines based on currently available information.

The occurrence of a hypothetical granuloblast is postulated (Fig. 18). From this cell arises young granulocytes, designated as progranulocytes. A progranulocyte has the following characteristics: (1) smallest of the circulating cells of the granulocyte line; (2) with relatively few cytoplasmic granules; (3) not actively but capable of phagocytosis; (4) basophilic nucleus and cytoplasmic granules; and (5) produce few pseudopodia when spread.

The progranulocyte gives rise to what is designated as a granulocyte I (Fig. 18). This cell is characterized by (1) being medium sized; (2) with numerous cytoplasmic granules some of which are basophilic, others are acidophilic, and still others are refractile; (3) being actively phagocytic; (4) with pseudopodia of filopodial type; (5) being capable of contributing to clumping; (6) with increased numbers of such organelles as Golgi, rough endoplasmic reticulum, liposomes and lysosomes; and (7) with relatively high levels of acid hydrolase activity.

The granulocyte I matures into a granulocyte II (Fig. 18). The latter is characterized by (1) being the largest of the granulocyte line; (2) including large numbers of cytoplasmic granules, mostly if not all are acidophilic; (3) including some cytoplasmic vacuoles, which are residues of the degranulation process; (4) being very actively phagocytic; (5) producing large numbers of semi-permanent filopodia when spread; (6) the presence of Golgi, smooth and rough endoplasmic reticulum, lipsosomes and lysosomes (granules of light microscopy in most species); (7) being actively involved in clumping; and by (8) having the highest levels of acid hydrolase activity. If a granulocyte II has phagocytosed foreign materials and intracellular degradation (digestion) has occurred, it becomes a spent granulocyte. Such a cell is characterized by (1) few filopodia, commonly at two poles; (2) few cytoplasmic granules; (3) larger number of cytoplasmic vacuoles of various shapes and sizes; (4) less lysosomal and hydrolase activity; (5) phagosomes commonly including digestive lamellae and amorphous, partially digested materials; and (6) clumps of glycogen granules present in the cytoplasm. Under certain pathologic conditions, such as during post-mortem change or rejection of incompatible grafts (Sparks and Pauley, 1964; Cheng and Galloway, 1970), several granulocyte IIs may become fused to become a multinuclear giant cell, designated as a macrocyte (Fig. 18). Not much beyond this is presently known about these cells.

What is presented above is this reviewer's current interpretation of the ontogeny and functional differentiation of the granulocyte line. There is a second line which is designated as the hyalinocyte series (Fig. 18).

As a granuloblast has been postulated for the granulocyte line, so a hypothetical hyalinoblast is proposed for the hyalinocyte line. It is postulated to be capable of dividing and differentiating into young hyalinocytes, designated as prohyalinocytes. Such a cell is characterized by (1) a relatively large nucleus surrounded by a small volume of cytoplasm; (2) the absence of cytoplasmic granules; (3) none or few lobopodia; and (4) being essentially basophilic.

The prohyalinocyte matures into a hyalinocyte (Fig. 18). This cell has (1) a relatively small volume of cytoplasm surrounding a large nucleus; (2) a few cytoplasmic granules; and (3) moves by producing a small number of lobopodia.

Finally, there is a third line comprised of serous cells, brown cells, or pigment cells (Fig. 18). Although, as reviewed later, these cells do occur in the general circulation, they are formed in Keber's glands. The youngest serous cells, i.e. those still within the gland or very recently expelled, are characterized by (1) being essentially non-motile; and (2) the presence of large light to dark brown pigment globules in the cytoplasm. The youngest serous cells are capable of dividing and they develop into medium serous cells. These include more pigment globules as well as acid mucopolysaccharide (Cheng and Burton, 1966). Young and large serous cells do not include acid mucopolysaccharide. The large serous cells are tightly packed with many pigment globules (Fig. 18).

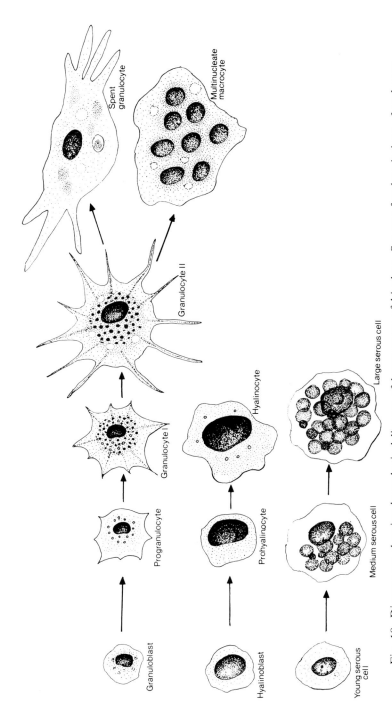

Spent
granulocyte

Multinucleate
macrocyte

Granulocyte II

Granulocyte I

Hyalinocyte

Large serous cell

Progranulocyte

Prohyalinocyte

Medium serous cell

Granuloblast

Hyalinoblast

Young serous
cell

Fig. 18. Diagramme showing hypothetical lineage of haemocytes of bivalves. See text for descriptions of each type.

V. Functions of haemocytes

Bivalve haemocytes are known to be involved in (A) wound repair, (B) shell repair, (C) nutrient digestion and transport, (D) excretion and (E) internal defence.

A. Wound repair

The role of bivalve haemocytes in wound repair has been expertly reviewed by Sparks (1972) and consequently need not be duplicated here in great detail.

Prior to commenting on wound repair, it appears appropriate to consider the clumping reaction as it occurs among bivalve haemocytes.

Drew (1910) studied the so-called clumping reaction in the haemocytes of *Cardium norvegicum*. He reported that the cells establish contact with one another by their filopodia, the interconnecting bridges then shorten and thicken, and as a result, contract, thereby drawing the cells together. A slight modification of this was reported by Dundee (1953) in several species of freshwater bivalves (Table I). The haemocytes of these bump into one another with their filopodia and subsequently form a mass. Also, Narain (1972a) reported that the haemocytes of *Lamellidens corrianus* become entangled by their filopodia and are eventually pulled together.

Foley and Cheng (1972) studied clumping by the haemocytes of *C. virginica*. They reported that it is the granulocyte that forms clumps; furthermore, it is only cells freshly drawn from the adductor muscle sinus that form aggregates which include from 3 to 4 up to 100–200 cells. However, if the preparation is permitted to stand, there is an exomigration of cells from each cluster in a more or less defined pattern and the fusion of abutting cells is not observed.

The clumping of bivalve haemocytes is physiologically different from the clotting of vertebrate blood. A fibrinogen-thrombin-fibrin system is absent in molluscs (Narain, 1972b).

Haemocyte clumping in bivalves is intimately involved in the wound healing process. Specifically, clumped haemocytes have been reported to delineate and plug wounds and arrest haemorrhage (Dakin, 1909; Drew, 1910; Goodrich, 1919; Orton, 1923; Des Voigne and Sparks, 1968, 1969; Ruddell, 1971b) (Figs 19, 20). These studies have been authoritatively reviewed by Sparks (1972); however, five significant aspects should be re-emphasized. (1) The initial phase of wound healing, at least in oysters, involves the infiltration of the wounded area by large numbers of haemo-cytes. (2) The aggregated haemocytes form a plug as well as delineating the

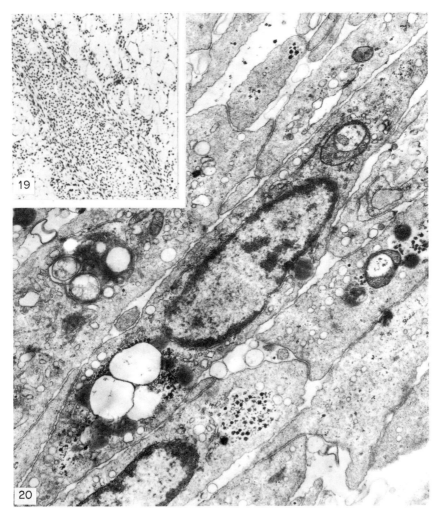

Fig. 19. Plug of haemocytes filling channel of a 94-h wound in *Crassostrea gigas*. Haematoxylin and eosin. (After Des Voigne and Sparks, 1968, with permission of Academic Press.) × 120.

Fig. 20. Electron micrograph of haemocytes of *Crassostrea gigas* comprising a plug in a 72-h wound. (After Ruddell, 1971b, with permission of Academic Press.) × 14 850.

area (Fig. 19). (3) Healing proceeds from the interior of the lesion toward the surface with the replacement cells, which are fusiform, having differentiated from the clumped haemocytes. In *C. gigas*, the wound channel is effectively plugged by 144 h post experimental wounding (Des Voigne and Sparks, 1968). (4) Collagen is deposited between the cells comprising the plug; however, in time, both the haemocytes and collagen are replaced by Leydig cells. (5) Phagocytic haemocytes, most probably granulocytes, eventually infiltrate and phagocytose the necrotic cell debris.

According to Des Voigne and Sparks (1968), the haemocytes originating from the original plug are totipotent and can differentiate into epitheloid tissue to replace the necrotic surface epithelium. Ruddell (1971a), however, disagreed and he believes that destroyed epithelial cells are replaced primarily, if not entirely, by the migration of adjacent epithelial cells. Support for Ruddell's contention is in the report by Mix and Sparks (1971) since during the healing of the digestive epithelium in *C. gigas*, following damage by ionizing radiation, the new epithelium results initially by mitotic division of remaining cells and subsequently by migration of adjacent healthy epithelial cells.

Ironically, it was Ruddell (1971a) who also reported that *C. gigas* haemocytes are capable of differentiating into other types of cells. Specifically, he reported that what he considered to be agranular amoebocytes, comprising plugs formed in response to experimentally inflicted mantle wounds (Fig. 20), may differentiate into fibroblasts and myoblasts. This report, if true, is supportive to the contention of Des Voigne and Sparks (1968) that haemocytes are not terminal cells. This reviewer doubts Ruddell's postulation because what Ruddell considered the differentiation from an "agranular amoebocyte" into a "myoblast" is probably the "differentiation" of granulocytes into spent granulocytes.

Pauley and Heaton (1969) studied wound repair in the freshwater mussel, *Anodonta oregonensis*, and four points deserve re-emphasis. (1) There was little cellular response associated with repair; rather, the edge of the wound invaginated in an apparent attempt to close the incision. This was often the only response noted during the first eight days. (2) Only infrequently was a clot formed in the wound channel and this resulted from infiltration and aggregation of haemocytes. The cells comprising the plug were spindle-shaped as in oysters, and these did not completely fill the wound channel. An extremely weak cellular response also occasionally developed in the surrounding tissues. (3) A few fibroblasts were intermingled initially with the haemocytes and these deposited scanty collagen-like material. Subsequently, the haemocytes were almost completely replaced with fibroblasts, which are connective tissue cells. (4) Replacement of the surface ciliated epithelium resulted from division and migration of adjacent healthy epithelial cells, and this occurred without the formation of a cellular wound plug. Thus, the lack of a pronounced initial cellular response and subsequent lack of involvement of large quantities of connective tissue fibres characterize wound healing in *A. oregonensis*.

The two most interesting aspects of the study by Pauley and Heaton from the standpoint of this review are that the wound plug was incomplete and essentially inconsequential, and the surface epithelium was replaced by division and migration of adjacent epithelial cells. Thus, there may be considerable truth to the statement by George and Ferguson (1950) that the clumping of haemocytes to form a plug is not as important in wound healing in molluscs as in vertebrates. The phenomenon, judging from Pauley and Heaton's experience, is incidental to the entire healing process and certainly does not appear to be of major haemostatic importance. Also, the results with *A. oregonensis* support Ruddell's (1971a) contention that haemocytes do not differentiate into epithelial cells.

B. Shell repair

A detailed review of shell repair among molluscs is unwarranted in view of the objectives of this contribution. Reviews by Wilbur (1964) and Abolinš-Krogis (1968) are available. Since the focus of this section is on the functional roles of haemocytes, only this aspect of the complex shell repair process is considered.

In brief, regeneration of molluscan shell involves (1) initiation of a stimulus or stimuli resulting from injury or experimental removal of a portion of the shell, (2) the mobilization of calcium and other substances from different regions of the organism, (3) transport of calcium and other materials to the area of repair, and (4) the localized deposition of the organic matrix and calcium carbonate (Sioli, 1935). The role of haemocytes in this series of events is in transport of calcium and protein.

Wagge (1951, 1955), who studied shell repair in the terrestrial gastropod, *Helix aspersa*, was the first to report that the transfer of calcium and organic shell matrix substances from the digestive gland to the site of repair is effected by haemocytes. Subsequently, Dunachie (1963) and Beedham (1965), who studied *M. edulis* and *Anodonta* sp., respectively, attributed the same function to the circulating haemocytes of these two bivalves. The chemical and biophysical bases of how molluscan haemocytes take up and transport calcium and protein is still uncertain. Intracellular phosphatase activity has been implicated, and, indeed, both acid and alkaline phosphatases are known to occur in these cells. However, intracellular phosphatase activity does not appear to be the entire story since Manigault

(1939) reported that the level of phosphatase activity in the serum of *Mytilus* sp. is increased during shell regeneration. Finally, Durning (1957) has questioned the importance of haemocytes in the transport of both calcium and proteins. As Wilbur (1964) has stated: "Even though amoebocytes may participate in repair processes, it would seem probable that the mantle contributes, directly and indirectly, the major share both of the calcium and the protein".

C. *Nutrient digestion and transport*

The role of haemocytes in nutrient digestion and transport, in view of recent findings, cannot be considered an isolated function. In fact, it is so intimately associated with internal defence that in many ways the two are indistinguishable. This has recently been reinforced by the results of Cheng and Rudo (1976a) and Cheng (1977). For this reason, details of intracellular digestion are considered under the subheading of "Internal defence". Only the general framework is considered here.

The role of haemocytes in digestion has been comprehensively reviewed by Takatsuki (1934a,b), Wagge (1955), Owen (1966), Purchon (1968) and Narain (1973), and need not be reconsidered in detail. Only some salient features are presented.

Digestion in bivalves is both extra- and intra-cellular. Extra-cellular digestion is effected in the stomach by enzymes released for dissolution of the crystalline style. Most investigators agree with the idea first promoted by Yonge (1937, 1946) that intracellular digestion is more important. This type of digestion occurs within two categories of cells: haemocytes and digestive cells of the digestive diverticula. In the first instance, it is recalled that bivalves have an open circulatory system and haemocytes occur not only within the heart, haemolymph vessels and sinuses but also are found migrating through tissues. Some migrate into the lumen of the alimentary tract from between the lining epithelium. These pinocytose soluble nutrients, some already partially digested by enzymes of style origin, and phagocytose particulate foodstuffs. Once taken into the cell, digestion commences, and concurrently, the haemocytes pass back into the deeper tissues of the body and in this manner transport nutrients to various tissues (Yonge, 1926).

In a most significant study, Feng *et al.* (1977) reported the finding of carotenoids, specifically flavenoids, β-carotene, canthaxanthin, and other unidentified xanthophylls, in haemocytes of *Mytilus coruscus* and *C. gigas.* The presence of these pigments, which can only be synthesized by plants, in bivalve haemocytes indicates that the molluscs must have acquired them by phagocytosing pigment-bearing unicellular algae which comprise the diet of these filter feeders. Furthermore, it is particularly significant that these investigators found carotenoids in haemocytes bled from the adductor muscle sinus. This strongly supports Yonge's (1926) contention that food particle-laden haemocytes in the lumen of the alimentary tract do traverse the lining epithelium and migrate into the deep tissues. Relative to this phenomenon, Feng *et al.* (1977) have raised an interesting point. Specifically, based on the results of Yonge (1926), Stauber (1950), Tripp (1960) and Feng *et al.* (1977), it can now be concluded that there is a two-way traffic of bivalve haemocytes across the epithelial lining of the alimentary tract and the movement of cells is not a random phenomenon. Therefore, an interesting question has to be asked as to what are the mechanisms governing the direction of haemocyte movement from the lumen to deep tissues and vice versa? Since Stauber (1950), Tripp (1960), and others reported that haemocytes laden with non-injurious foreign materials, e.g. India ink particles, bacterial spores, etc. migrate through the alimentary wall to the lumen, and Yonge (1926) and Feng *et al.* (1977) found the migration of cells laden with materials of food origin from the lumen through the epithelium into deep tissues, it may be inferred that the directional migration of haemocytes is in some way directed by their exogenous inclusions.

The second site of intracellular digestion in bivalves is the digestive diverticula. The end products of digestion within digestive cells of the diverticula are believed to be transferred to haemocytes and serum, which serve to transport them to the various tissues of the body.

D. Excretion

Although the physiology of the excretory system of bivalves has been studied extensively (see reviews by Grassé, 1960; Galtsoff, 1964; Martin and Harrison, 1966), relatively little is known about the role of haemocytes in this process. Cheng *et al.* (1969), however, have demonstrated that India ink experimentally injected into the marine prosobranch, *Littorina scabra*, is rapidly phagocytosed by haemocytes, most of which are subsequently eliminated from the body via exodus through the epithelia of the nephridial tubules. Furthermore, although Durham (1891) reported that degenerated pigments are excreted from *Anodonta* sp. and *Unio* sp. via the exodus of laden haemocytes, Canegallo (1924) described the excretion of oil globules in *Unio* sp. in the same way, and Orton (1923) also reported the elimination of metals from oysters via this mechanism, it was, however, not ascertained whether the nephridial tubules were the main site of excretion. This also holds true for the removal of necrotic haemocytes (Sparks and Pauley, 1964), Thorotrast (Nakahara and Bevelander, 1969), and ceroid-like pigments (Zacks, 1955) from bivalves.

Another aspect of the excretory function of haemocytes involves the pericardial glands, which are also known as Keber's glands. These glands, which comprise a part of the excretory system, are prominent in most bivalves, occurring in two locations; on the auricles and/or in the mantle, and empty into the pericardial cavity.

The Keber's glands are also the sites for the production of the so-called brown or serous cells. These pigment-bearing cells occur not only in tissues but sometimes also within the closed portions of the circulatory system, and hence comprise a type of circulating and migrating cell. Their colouration is due to yellow to brown cytoplasmic globules. One of the most detailed descriptions of brown cells is that given by Takatsuki (1934b) who reported that they are amoeboid and capable of phagocytosing carmine particles. He proposed that they are modified leucocytes.

White (1942) reviewed the possible excretory function of serous cells. Basing his hypothesis on the finding of hippuric acid as the main component of the cytoplasmic globules (or granules) associated with the pericardium of *Cardium* sp. and *Pecten* sp. by Letellier (1891), and the occurrence of hippuric acid and other metabolic by-products in the pericardial fluid of *Ostrea edulis* by Takatsuki (1934b), White proposed that serous cells are excretory in function and extract acids from the blood which they carry to the kidney (organ of Bojanus) via the ciliated renopericardial channels. In addition to this assumed function, storage of fats and the secretion of shell-forming materials are other proposed functions (see White, 1942). Furthermore, they are believed to protect the musculature of the auricles (Takatsuki, 1934b) and are in some manner correlated with ageing (Turchini, 1923).

The most important function attributed to serous cells is related to their role in the removal of degradation products of dead or moribund parasites and the metabolic by-products of successful parasites. Relative to this role, several investigators have noted increased pigmentation in *C. virginica* and other oysters infested with the annelid, *Polydora* spp., the oyster crab, *Pinotheres ostreum*, and *C. virginica* parasitized by the sporozoan *Haplosporidium* (= *Minchinia*) *nelsoni*. The first published account of the possible functional association of brown cells with parasitism is that by Mackin (1951) who noted an increase in the number of these cells in *C. virginica* parasitized by the fungus *Perkinsus marinus* (= *Labyrinthomyxa marina* and *Dermocystidium marinum*). He postulated that the increase may be indicative of an inbalance in lipid metabolism, possibly due to an increase in the oxidative and reductive processes resulting from disease. Mackin (1951) also noted an increase in the number of serous cells in *Macoma balthica* parasitized by *Perkinsus* sp. and is of the opinion that the relative resistance of this clam to the fungus is the result of the large number of serous cells produced. Mackin's (1951) earlier observations have been strengthened by the quantitative studies by Stein and Mackin (1955). Similarly, Cheng and Burton (1965, 1966) noted that although serous cells are present in non-parasitized *C. virginica*, their number is increased in oysters parasitized by sporocysts of the trematode *Bucephalus* sp. Serous cells found in the tissues of *C. virginica* vary greatly in size. Each cell may measure from 4 to 18 μm in diameter. Their cytoplasm is usually so packed with yellowish brown to dark brown globules that the nucleus is normally masked. These cells have been observed to undergo division in oyster tissue (Cheng and Burton, 1966).

Haigler (1964) reported the presence of "granules" of lipid and tyrosine rich protein in the serous cells of *C. virginica*. Furthermore, as the result of the observations by Stein and Mackin (1955) and Haigler (1964), the granules are known to possess properties similar to those of lipofuchsins although, as Haigler pointed out, the "lipofuchsin" present in oyster serous cells is soluble in dilute acids and bases. It is noted that Stein and Mackin (1955) reported that the serous cells in the same species of oyster are insoluble in dilute acids and bases as one would expect of true lipofuchsins. Lipofuchsins represent a class of lipogenous pigments derived from lipoid or lipoprotein sources (Pearse, 1961) or fatty acids (Gomori, 1952; Lillie, 1954).

The complexity of the chemical composition of the globules in serous cells is further indicated by Cheng and Burton (1966) who found that only some of these cells in *C virginica* are periodic acid-Schiff (PAS) positive and diastase resistant. Thus, the chemical nature of this material could be mucoprotein, a glycoprotein, a glycolipid, or a sphingolipid. Since Haigler (1964) has shown that serous cells are composed of a complex of lipid and protein with properties similar to those of lipofuchsins, it would appear that the PAS-positive and diastase resistant material can be classified as sphingomyelin and the PAS-positive reaction is not due to the presence of a 1:2-glycol containing carbohydrate but to the presence of a primary acylated amine adjacent to a hydroxyl group that is capable of reacting with periodic acid. In addition to sphingomyelins, Cheng and Burton (1966) also demonstrated the presence of acid mucopolysaccharides in serous cells of medium size (6–12 μm). In view of the above, if serous cells are to be considered a specialized type of haemocyte, then the role of haemocytes in excretion is well established.

E. Internal defence

As stated in the "Introduction", most of the research emphasis in recent years on bivalve haemocytes has been directed to understanding their roles in internal defence against nonself materials.

Stauber (1961), in his review of immune mechanisms in invertebrates, especially the American oyster, *C. virginica*, informally defined cellular defence as being of three categories: phagocytosis, encapsulation and leucocytosis. This scheme has since been adopted by several authors (Cheng and Sanders, 1962; Feng, 1967; Cheng, 1967; and others) and extended to include nacrezation.

1. Phagocytosis

Phagocytosis is a well-known type of internal defence mechanism in invertebrates as well as vertebrates. It involves the uptake of foreign materials by certain types of host cells and thus prevents direct contact of such materials, biotic or abiotic, with the host's tissues. The fate of phagocytosed materials may differ. Modern studies on molluscan phago-cytosis commenced with Stauber (1950) who traced the ultimate disposition of India ink particles experimentally introduced into *C. virginica*. Since then, critical reviews of the fates of a variety of experimentally introduced foreign materials which are phagocytosed in molluscs have been published (Cheng, 1967; Feng, 1967; and others). Briefly, it is now known that digestible particles and macromolecules are degraded within oyster leucocytes (Tripp, 1958a,b, 1960; Feng, 1959, 1965) while indigestible particles and macro-molecules are voided via the migration of foreign material-laden phagocytes across certain epithelial borders (Stauber, 1950; Tripp, 1960; Feng, 1965) (see section on excretion, above).

It is noted that not all phagocytosed foreign materials are eliminated by the mechanisms mentioned. Some are retained within the cytoplasm of host cells for relatively long periods while others, specifically, a number of microorganisms (certain bacteria and the fungus, *Perkinsus marinus*), can grow and multiply therein (Prytherch, 1940; Mackin, 1951; Michelson, 1961). Some of these intracellularly sustained microorganisms have, through evolutionary adaptation, become mutualists of their molluscan hosts. The most striking example of this is the occurrence of intracellular zooxanthellae in the marine bivalves *Hippopus* sp. and *Tridacna* sp. (see Yonge, 1936) and in the nudibranch *Aeolidiella* sp. (see Naville, 1926).

Finally, it is noted that not all invading organisms are phagocytosed. It has been reported that the presence of the sporozoan, *Haplosporidium nelsoni* and the injection of the flagellate, *Hexamita nelsoni* and of

Staphylococcus aureus phage 80 into *C. virginica* induces little or no phagocytosis (Feng, 1966; Canzonier, quoted in Feng, 1967; Feng and Stauber, 1968). Similarly, Goetsch and Scheuring (1926), Yonge and Nicholas (1940) and Buchner (1965) reported that the mutualistic zoo-chlorellae in *Lymnaea* sp., *Anodonta* sp. and *Unio* sp. are seldom found within host cells and are presumed to lead an extracellular existence.

(*a*) *Cell types involved.* That molluscan haemocytes are capable of phagocytosis has been known since Haeckel's (1862) report on *Helix* sp. and *Thetis* sp.; however, it was uncertain which cell type was the most active from this standpoint. Although most investigators believed that granu-locytes were most active, it was not until Foley and Cheng (1975) performed a quantitative study on the haemocytes of *C. virginica* and *M. mercenaria* that it was ascertained that all of the cells are phagocytic; however, of these the granulocytes were considerably more active. When the percentage of what was then termed "primary fibrocytes" and "secondary fibrocytes" of *C. virginica*, which are now known to be degranulated granulocytes, that had phagocytosed bacteria are added to the percentage of granulocytes that had engulfed bacteria, it was ascertained that 87·28% of granulocytes as compared to 12·32% of hyalinocytes were associated with experimentally introduced *Staphylococcus aureus*, and 83·48% of granulocytes as compared to 16·80% of hyalinocytes were associated with experimentally introduced *Escherichia coli*. In the case of haemocytes of *M. mercenaria*, 12·82% of granulocytes as compared to 2·56% of hyalinocytes were associated with bacteria.

Also, Foley and Cheng (1975) demonstrated that if the haemocytes were exposed to bacteria *in vitro* at 4, 22 and 37°C, the phagocytic indices were higher at the two higher temperatures. This finding is in general accord with *in vivo* experiments on clearance rates performed by others (Table II).

(*b*) *Uptake mechanisms.* Bang (1961) appears to be the first to report on detailed observations of how foreign particles are endocytosed by bivalve haemocytes. He found that motile bacteria about to be phagocytosed by granulocytes of *C. virginica* initially stick to the surface of the molluscan cell, commonly to the surfaces of the filopodia (Fig. 21). Subsequently, they are taken into the ectoplasm (Fig. 21) by gliding along filopodia and become enclosed in a phagosome.

A second uptake mechanism was reported by Cheng (1975). As a result of contact between bacteria and granulocytes of *C. virginica*, invaginations of the cell surface develop and the bacteria are taken up into endocytotic vacuoles (Fig. 22). No filopodia are involved.

A third uptake mechanism was recently reported by Renwrantz *et al.* (1979). They studied the uptake of rat erythrocytes by *C. virginica* haemocytes *in vitro* and reported that both granulocytes and hyalinocytes,

TABLE II. A comparison of the results of clearance experiments using *Crassostrea virginica* during which the effects of different temperatures were studied.

Foreign material or organism employed	Temperature (°C)	Duration of experiment	Amount of inoculum	Result	Reference
Bacteriophage 80	5	140 h	$0.2 \times 10 \log_{10}$ plaque-forming units	43 h required to reduce by 1 $\log_{10}$ PFU[a]/ml	J. S. Feng (1966)
Bacteriophage 80	15	4 days	$0.2 \times 10 \log_{10}$ plaque-forming units	3–4 h required to reduce by 1 $\log_{10}$ PFU/ml	J. S. Feng (1966)
Bacteriophage 80	23.5	4 days	$0.2 \times 10 \log_{10}$ plaque-forming units	1.5 h required to reduce by 1 $\log_{10}$ PFU/ml	J. S. Feng (1966)
Pseudomonas-like, A-3	9	35 days	2×10^7	steady decline in number	S. Y. Feng (1966)
Pseudomonas-like, A-3	16	35 days	2×10^7	decrease in number for 4 days, increase in number for 4 days, then steady decline in number	S. Y. Feng (1966)
Pseudomonas-like, A-3	23	20 days	2×10^7	2 fluctuations in number of organisms in 20 days	S. Y. Feng (1966)
Bacteriophage T2	25	48 h	$1 \times 10^{10} - 8 \times 10^{10}$ plaque-forming units	48 h required to reduce 3 $\log_{10}$ PFU/ml	Acton and Evans (1968)
Bacteriophage T2	32–34	24 h	$1 \times 10^{10} - 8 \times 10^{10}$ plaque-forming units	24 h required to reduce < 1 $\log_{10}$ PFU/ml	Acton and Evans (1968)
Hexamita sp.	6	18 days	2×10^7	100% oyster mortality within 18 days	Feng and Stauber (1968)
Hexamita sp.	12	20 days	4×10^5	80% oyster mortality within 14 days	Feng and Stauber (1968)
Hexamita sp.	12	15 days	1×10^3	50% oyster mortality within 14 days	Feng and Stauber (1968)
Hexamita sp.	18	14 days	8.8×10^5	no patent infection; 0% oyster mortality within 14 days. Few phagocytosed Hexamita observed	Feng and Stauber (1968)
Hexamita sp.	18	14 days	4×10^5		
Hexamita sp.	18	14 days	1×10^3		
Chicken erythrocytes	6	30 days	$1.1 \times 10^5 - 2.62 \times 10^5$	extensive phagocytosis, small number of non-phagocytosed RBC's present	Feng and Feng (1974)
Chicken erythrocytes	15–19	22 days	$1.1 \times 10^5 - 2.62 \times 10^5$	RBC's digested within phagocytes or removed by exomigration of RBC-laden phagocytes	Feng and Feng (1974)

[a] Plaque-forming units.

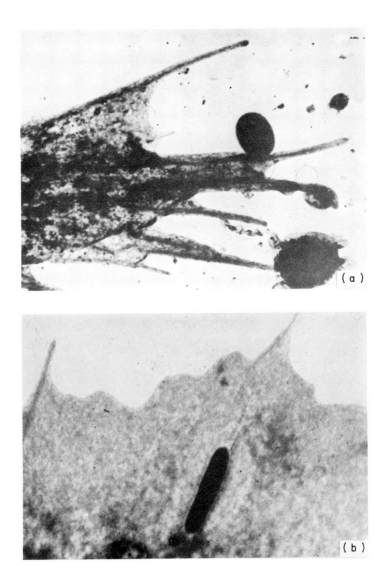

Fig. 21. Uptake of bacteria by granulocyte of *Crassostrea virginica*. (a) Adherence of bacterium to filopodium. × 13 500. (b) Subsequent endocytosis of bacterium. (After Bang, 1961. With permission of *Biological Bulletin*.) × 8600.

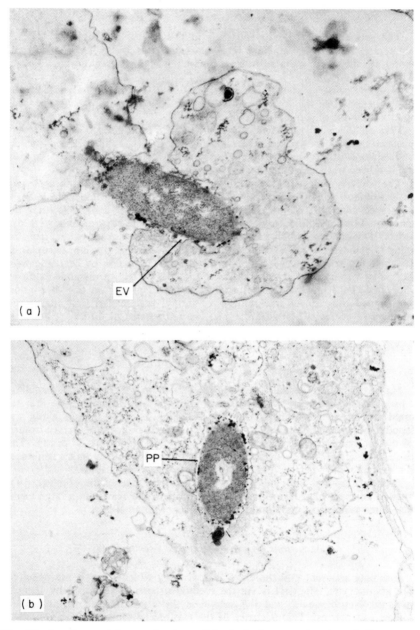

Fig. 22. Another uptake mechanism of bacteria by granulocytes of *Crassostrea virginica*. (a) Engulfment of *Bacillus megaterium* by molluscan cell without the aid of filopodium. ×20 000. (b) Subsequent endocytosis of bacterium into primary phagosome (PP). ×20 000. (After Cheng, 1975, with permission of New York Academy of Sciences.)

primarily the former, commonly take in red blood cells by producing a funnel-like pseudopod through which the foreign cell glides into a phagosome in the ectoplasm (Fig. 23). This was the first report of this uptake mechanism in any invertebrate. It is now known to occur with granulocytes of the gastropods *Biomphalaria glabrata* and *Bulinus truncatus* (see Schoenberg and Cheng, 1980).

Fig. 23. Third uptake mechanism by granulocytes of *Crassostrea virginica*. (A) Beginning of uptake of rat erythrocyte through "funnel" formed by granulocyte. (B) Subsequent endocytosis of erythrocyte.

(*c*) *Intracellular degradation*. The processes involved in intracellular degradation within bivalve haemocytes have been somewhat clarified as a result of electron microscope and biochemical studies. The process is apparently the same whether it is foodstuffs or engulfed foreign molecules or organisms that are being digested. Since granulocytes are more phagocytic than hyalinocytes (Foley and Cheng, 1975), and one of the major differences between these types of cells is the occurrence of large numbers of cytoplasmic granules in granulocytes, it became evident that the nature of these granules had to be resolved. In the case of granulocytes of *M. mercenaria*, Yoshino and Cheng (1976a) have demonstrated by employing cytochemistry coupled with electron microscopy that these granules are true lysosomes. They serve as storage organelles for acid hydrolases, and are therefore analogous to the granules in mammalian polymorphonuclear and monocytic leucocytes (see Chapter 17).

Studies on several species of bivalves, i.e. *M. mercenaria*, *Mya arenaria*, and *Spisula solidissima* (Cheng, unpublished), have revealed a similar pattern of events (Fig. 24). Eventually, after enzyme digestion is completed, certain molecules, such as monosaccharides and fatty acids, apparently diffuse through the phagosomal membrane into the cytoplasm. At this site, glucose is synthesized into glycogen and numerous glycogen granules aggregate in the cytoplasm. Indigestible materials remain within the original phagosomes, which are now residual bodies (or vesicles), and are eventually discharged from the cell (Fig. 24).

It is noted that not all lysosomes are associated with intracellular digestion. As reviewed in the subsection below, external contact with certain foreign substances will result in hypersynthesis of intracellular lysosomal enzymes which are released from haemocytes into the serum where digestion of the foreign material, such as bacteria, is initiated. The release of enzymes is effected by what has been termed degranulation (Foley and Cheng, 1977), a process involving the migration of lysosomes (cytoplasmic granules) to the surface of the cell where the enclosed enzymes are discharged.

The intracellular digestion process in oysters is somewhat different from that in the other bivalves which have been studied. Details of this process have been reviewed by Cheng (1975) and are illustrated in Fig. 25. In time, this digestive process results in the formation of digestive lamellae around the degraded bacteria (Figs 25, 26) and the accumulation of large numbers of glycogen granules, synthesized from glucose, within secondary phago-somes (Figs 25, 27). Eventually, the secondary phagosomal wall dis-integrates and the glycogen granules are freed in the cytoplasm (Fig. 28). These are then discharged from the cell into the serum in membrane-bound

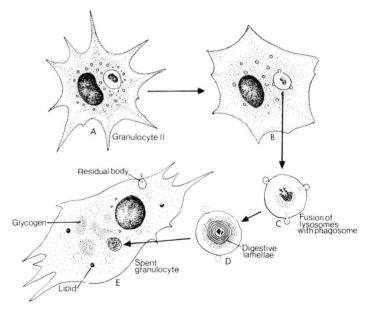

Fig. 24. Sequence of events during the intracellular degradation of foreign material in granulocyte of *Mercenaria mercenaria*. (A) Foreign material in phagosome. (B–C) Migration and fusion of lysosomes to phagosome. (D) Formation of digestive lamellae around partially degraded foreign material. (E) Spent granulocyte with phagosome enclosing digestive lamellae, lipid droplets, aggregates of glycogen granules in cytoplasm and residual body.

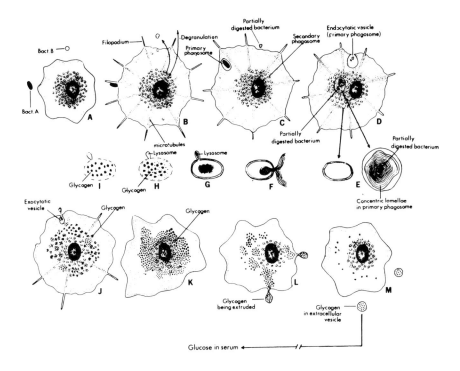

Fig. 25. Sequence of events that occur during the phagocytosis and intracellular degradation of bacteria by granulocytes of *Crassostrea virginica*. (A) Granulocyte in presence of bacteria A and B. (B) Bacterium A becomes attached to filopod, while bacterium B is altered by enzyme(s) release from granulocyte during degranulation. (C) Bacterium A is within primary phagosome of granulocyte. (D) Bacterium B is taken into granulocyte by endocytosis and digestion of bacterium A has commenced within primary phagosome. (E) Formation of digestive lamellae around bacterium in primary phagosome; secondary phagosome in vicinity. (F) Transfer of partially digested bacterium from primary to secondary phagosome. (G) Fusion of lysosomes with secondary phagosome. (H) Glycogen synthesized from sugar constituents of degraded bacterium; phagosomal wall disintegrating. (I) Phagosomal wall disintegrating. (J) Discharge of nondegradable remnants of bacterium via exocytotic vesicle; accumulation of glycogen in cytoplasm of granulocyte. (K) Massing of glycogen in cell and disappearance of filopodia. (L) Glycogen in process of being discharged into serum in packets. (M) Glycogen discharged.

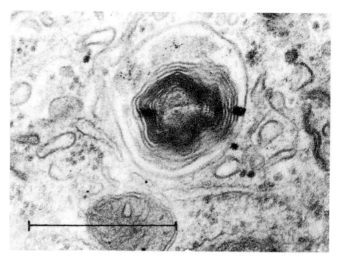

Fig. 26. Electron micrograph showing formation of digestive lamellae around degraded bacterium in secondary phagosome of *Crassostrea virginica* granulocyte. Bar = 1 μm.

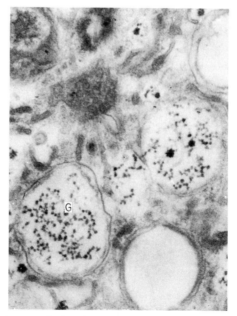

Fig. 27. Electron micrograph showing accumulation of glycogen granules (G) within secondary phagosomes of *Crassostrea virginica* granulocytes. × 36 800.

vesicles. These are subsequently degraded in serum, presumably by released lysosomal enzymes, and the glycogen is by some pathway degraded to glucose and employed as an energy source.

The mechanism in oysters summarized above has been confirmed by Cheng and Rudo (1976a) who injected ^{14}C-labelled *Bacillus megaterium* into the American oyster, *C. virginica*, and subsequently extracted glycogen from their haemocytes, sera, and body tissues at several time intervals. It was found the ^{14}C was first detected in glycogen extracted from haemocytes and body tissues, and later from serum.

(*d*) *Energy requirements*. Since endocytosis, intracellular digestion, and a number of associated processes are energy-requiring ones, Cheng (1976c) conducted a study to ascertain the source of the energy. In brief, it was found that unlike mammalian phagocytes, there is no increase in O_2 consumption by haemocytes of *M. mercenaria* actively phagocytosing *Bacillus megaterium*. The utilization of glucose and glycogen, coupled with the production of lactate and no increase in O_2 uptake, indicate that glycolysis is the energy-providing pathway. This conclusion is strengthened by the fact that KCN does not inhibit phagocytosis. It was also found that

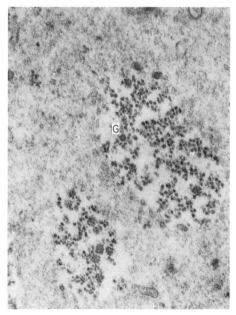

Fig. 28. Electron micrograph showing clumps of glycogen granules (G) free in the cytoplasm of a spent granulocyte of *Crassostrea virginica*. × 36 800.

nitro-blue tetrazolium reduction characteristic of mammalian phagocytes is absent in *M. mercenaria* haemocytes, and the myeloperoxidase-H_2O_2-halide antimicrobial system of mammalian phagocytes is also wanting (see Chapter 17).

2. Encapsulation

Encapsulation involves the enveloping of an invading organism or experimentally introduced tissue too large to be phagocytosed by host cells. Although much is known about the nature of such capsules as the result of descriptive studies, relatively little is known about the dynamics of the process. Furthermore, several types of encapsulation have been proposed, some involving cells and others involving connective tissue fibres (Cheng and Rifkin, 1970; Harris, 1975).

To date, almost all that is known about encapsulation has stemmed from studies on the reaction of bivalves to helminth parasites. This topic has been reviewed by Cheng and Rifkin (1970) and little new information has been contributed since then.

3. Some common denominators

Cheng and Rifkin (1970) proposed that encapsulation represents an abortive attempt at phagocytosis. This concept is based on the observation that the first cells to approach a foreign body too large to be phagocytosed become intimately flattened against its surface in a fashion which could be interpreted as an unsuccessful attempt at phagocytosis. Thus, in addition to chemotaxis, contact or attachment between the phagocyte and the foreign body are common denominators of phagocytosis and encapsulation. There is considerable evidence for chemotaxis towards certain bacteria and metazoan parasites (Cheng *et al.*, 1974; Cheng and Rudo, 1976b; Cheng and Howland, 1979) (see section on leucocytosis, below).

Relative to attachment, although Renwrantz and Cheng (1977), by employing agglutinins with known sugar specificities, have demonstrated that there are specific binding sites on haemocytes of the gastropod, *Helix pomatia*, such studies have yet to be conducted on bivalve haemocytes. The only exception is the study by Yoshino *et al.* (1979) who investigated the binding and redistribution of concanavalin A (Con A) surface membrane determinants in two sub-populations of *C. virginica* granulocytes as ascertained by size and designated as LS (5–9 µm) and SS (2–5 µm). By employing a direct Con A-peroxidase coupling method in conjunction with light and electron microscopical cytochemistry, it was determined that in a phosphate buffered saline medium, polar redistribution or capping of Con A receptors was observed in 20·4% of LS and 9·9% of SS granulocytes. Sodium azide (5×10^{-2} M) inhibited LS and SS granulocyte cap formation by 49 and 58%, respectively. Inhibition was $-N_3$ concentration-dependent. At the lowest concentration of $-N_3$ tested (5×10^{-4} M), removal of the inhibitor resulted in a complete reversal of capping inhibition for LS, but not SS cells. Also, when an isotonic saline medium (0·27 M NaCl) was substituted for phosphate buffered saline in all reagents, including washing solutions, the number of capforming cells almost doubled in both subpopulations of haemocytes. Preincubation of cells in a "greater than 1000 MW" homologous serum fraction suppressed capping of Con A receptors back to levels comparable to the phosphate buffered saline control.

This study by Yoshino et al. (1979) pointed out that there is at least a mannose-specific binding site on the oyster granulocytes studied since Con A is specific for this sugar. Also, it demonstrated that there are functional differences between LS and SS subpopulations of C. virginica granulocytes. These, according to the classification presented earlier (Feng et al., 1977) most probably reflect age differences.

The immunologic significance of capping is still unclear. In the case of vertebrate cells it has been suggested that surface receptor mobility and aggregation may function in antigenic modulation, immune expression (in lymphocytes) or inter- and intra-cellular recognition and communication (see Singer, 1974, and Edelman, 1976, for reviews).

4. Leucocytosis

Leucocytosis, which is defined as an increase in the number of leucocytes or haemocytes, is a forerunner of phagocytosis and/or encapsulation since the increased number of cells contributes to these active processes. The present state of our knowledge relative to leucocytosis as associated with parasitism, including helminth parasitism, has been conveniently summarized by Farley (1968) in tabular form. There is one aspect of leucocytosis which deserves come comment. Since it still remains uncertain where haemocytes of bivalves are formed, it cannot be stated with certainty that true leucocytosis occurs, i.e. although the occurrence of large aggregates of haemocytes surrounding protozoan and metazoan parasites and/or other types of foreign bodies is well known, it remains undetermined whether these cells are newly formed at some haemopoietic centre(s) or represent old cells which are released from sequestration in tissues. It is known that chemotaxis occurs between oyster haemocytes and metazoan parasites and certain live bacteria (Cheng et al., 1974; Cheng and Rudo, 1976b; Cheng and Howland, 1979); however, neither the chemotactic molecule(s) nor whether the cells responding to the foreign agent are formed de novo are known. Similarly, although Cheng (1966a,b) has reported a somewhat unique type of leucocytosis involving large numbers of haemocytes surrounding haemolymph vessels in which are located the nematode. *Angiostrongylus cantonensis*, it is not known whether these cells were formed *de novo* or represent cells released from sequestration sites in tissues.

5. Nacrezation

Nacrezation is a term coined by Cheng (1967) to describe the deposition of nacre around parasites of molluscs which irritate or invade the mantle region. The result is pearl formation. Pearl formation within certain pelecypods has been known for over 1000 years. According to Tsujii (1960), pearl culture was recorded in the *Gokango* by Hanyo, a Chinese, in about 80 A.D. The cause and mechanisms involved, however, were not understood until the nineteenth century, although it has been recognized as a type of defence mechanism against zooparasites since the report in 1655 by Worm that pearly formations occur in the mantle of *Mytilus edulis* collected in Sweden. Since this original report, several twentieth century biologists have investigated this phenomenon (Dubois, 1901, 1907; Jameson, 1902; Perrier, 1903; Giard, 1907; and others). It is known that certain trematode metacercariae, especially those of *Meiogymnophallus minutes* (= *Gymnophallus margaritarum*), when found between the inner surface of the shell and the mantle of marine pelecypods, will stimulate the mantle to secrete nacre which becomes deposited around the parasites. The process has been reviewed by Alverdes (1913). It does not usually involve haemocytes.

6. Haemocyte-associated humoral factors

When one considers resistance to parasites and other types of non-self materials in vertebrates, a distinction can be conveniently made between innate and acquired mechanisms. Although invertebrate immunologists have also categorized resistance as being innate or acquired, these should not be confused with their counterparts in vertebrates as there are fundamental differences. For example, it is now widely accepted among comparative immunologists that invertebrates, including molluscs, do not synthesize immunoglobulins nor have the classical complement system; hence if acquired resistance does occur among invertebrates, it is not based on antigen-antibody interaction.

There are some reports available which suggest the presence of innate humoral factors in molluscs. For example, Tripp (1960) reported that the haemolymph of the American oyster, *C. virginica*, kills certain bacteria.

Bacteria inoculated into this bivalve are rapidly destroyed extracellularly. Cheng *et al.* (1966) reported the occurrence of another type of innate humoral substance in bivalves. Specifically, they found that the serum of several species of marine pelecypods, especially that of *C. gigas*, will rapidly induce the ectopic immobilization and encystment of the cercariae of the echinostome, *Himasthla quissetensis*, and thus prevent their penetration into the second intermediate host.

The question that has been asked is: do both innate and acquired humoral factors exist in bivalves as suggested by the indirect evidence available? If so, what is the nature of these factors? Summarized at this point is our knowledge relative to humoral factors in bivalves.

Since cellular reactions to non-self materials in the form of phagocytosis and encapsulation are known to be the primary forms of internal defence mechanisms in bivalves (Stauber, 1950; Tripp, 1960; Feng, 1965; Cheng *et al.*, 1969; Cheng and Rifkin, 1970; and others), and since morphological evidence for intracellular degradation of phagocytosed materials is available, initially we were interested in determining what enzymes occur within molluscan phagocytes and what are their biochemical and biophysical characteristics (Rodrick and Cheng, 1974b; Cheng and Rodrick, 1974; Cheng, 1976a).

As a result of finding lysosomes in granulocytes of different species of bivalves (Cheng, 1975; Yoshino and Cheng, 1976a), and discovering that these organelles are associated with intracellular degradation (Cheng and Cali, 1974; Cheng *et al.*, 1974), identification of the lysosomal enzymes was conducted. Thus, Cheng and Rodrick (1975) reported that β-glucuronidase, acid phosphatase, alkaline phosphatase, lipase, aminopeptidase and lysozyme are associated with both the cellular and serum components of *C. virginica* and the hard clam, *M. mercenaria*.

The finding of these lysosomal enzymes in both cells and serum of bivalves led to the investigation of their possible origin in serum. Cheng *et al.* (1975) demonstrated that when haemocytes of *M. mercenaria* are actively phagocytosing *Bacillus megaterium in vitro*, lysozyme is released from cells into serum, and Cheng and Yoshino (1976) demonstrated that there is an increase in lipase activity in both the haemocytes and serum of the soft-shell clam, *Mya arenaria*, during *in vivo* phagocytosis of heat-killed *B. megaterium*.

Yoshino and Cheng (1976b) studied alterations in the levels of aminopeptidase activity in both the haemocytes and serum of *C. virginica* after *in vitro* exposure to heat-killed *B. megaterium*. They reported that there was a significant elevation in intracellular aminopeptidase activity induced by the challenge; however, there was no increase in the activity of this enzyme in the serum. These data suggest that aminopeptidase is not released into serum by haemocytes of *C. virginica* during phagocytosis although the bacterial challenge did stimulate hypersynthesis of this enzyme within haemocytes. Possibly, the release of aminopeptidase from molluscan haemocytes into serum may only be induceable by certain types of antigenic challenge and not by others. Whether this phenomenon holds for other lysosomal enzymes remains to be tested.

How then are lysosomal enzymes released into serum? Foley and Cheng (1977), by employing granulocytes of *M. mercenaria* as the model, demonstrated that degranulation occurs during phagocytosis and this process represents the morphological basis for enzyme release from the cytoplasmic granules, which are true lysosomes (Yoshino and Cheng, 1976a).

Having established that the overall pattern of enzyme activity during phagocytosis involves stimulation by certain antigens which results in the hypersynthesis of lysosomal enzymes and their subsequent release by the process of degranulation into serum, the question had to be asked as to whether the serum enzymes serve a function relative to internal defence?

To date, only the antimicrobial property of one lysosomal enzyme, lysozyme, is known. McDade and Tripp (1967) reported that this hydrolase from the haemolymph of the oyster, *C. virginica*, will lyse certain Gram-positive bacteria, specifically *Micrococcus lysodeikticus*, *Bacillus megaterium* and *Bacillus subtilis*. In addition, Rodrick and Cheng (1974b) demonstrated that the lysozyme of *C. virginica* will lyse *Escherichia coli*, *Gaffkya tetragena*, *Salmonella pullorum* and *Shigella sonnei*, in addition to *B. subtilis* and *B. megaterium*. It has no effect on *Staphylococcus aureus*. Also, Cheng and Rodrick (1974) reported that the lysozyme of *Mya arenaria* will lyse *M. lysodeikticus*, *B. megaterium*, *Proteus vulgaris*, *S. pullorum*, *S. sonnei*, *B. subtilis* and *E. coli*, but also has no effect on *S. aureus*. Such information has been obtained by assaying for the breakdown products of these bacteria.

There is little doubt that the destruction of bacteria by induced elevations in the levels of lysosomal enzymes is not as specific as that resulting from immunogen-immunoglobulin interaction. However, the surface of the exposed bacterium must contain a chemically compatible substrate.

As a result of the studies by McDade and Tripp (1967), Rodrick and Cheng (1974a), and Cheng and Rodrick (1974), a pattern is evolving, i.e. elevated levels of lysosomal enzymes in serum can serve as defence molecules against susceptible microorganisms. However, it is still not completely understood why certain invading organisms are destroyed while others are not. Furthermore, what is the basis for the encapsulation and subsequent destruction of certain strains of metazoan parasites in incompatible strains of bivalves?

One postulation as to why elevated serum hydrolases do not affect certain non-self substances is the absence of vulnerable substrates on their surfaces.

The second possibility is that the release of lysosomal enzymes from their sites of synthesis, i.e. granulocytes and other sources (Yoshino and Cheng, 1977), must be triggered by some component of the parasite's somatic antigens and/or secretions. If such a triggering molecule(s) is absent, then one would not expect the release of enzymes, at least to deleterious levels. The third possibility is that anti-enzymes are elaborated by the parasite which inactivates the lytic enzymes. This remains to be investigated.

In summary, there is now a body of evidence that there are acquired "humoral" protective molecules in bivalves but these are not immunoglobulins or opsonins. These are lysosomal enzymes released primarily from granulocytes and which are limited to their specificities. It is possible that these molecules are responsible, at least in part, for the anti-microbial properties of molluscan sera and extracts that have been reported.

It needs to be emphasized that lysosomal enzymes do occur at a lower level in the serum naive bivalves. Apparently, the levels in such molluscs are insufficient to prevent the invasion and establishment of compatible prokaryote and eukaryote parasites.

Finally, it bears re-emphasizing that degradation of foreign substances in haemocytes and/or serum of bivalves, based on presently available information, results in acquisition of nutrients. For this reason, I am of the opinion that from the evolutionary viewpoint, the original functions of molluscan phagocytes were nutrient acquisition, digestion and circulation, and that it was only later that these cells adopted their important role in internal defence.

VI. Summary and concluding remarks

It is proposed that for the sake of uniformity, the haemocytes of bivalves be designated at (1) granulocytes, (2) hyalinocytes and (3) serous cells. Furthermore, the designations employed in Fig. 18 for the ontogenetic stages of these three categories of cells should be given serious consideration for uniform use.

Numerous studies still remain to be conducted before a more complete understanding of bivalve haemocytes can be obtained. Among these are: (1) determination of the haemopoetic sites and mechanisms; (2) detailed studies on the structural and biochemical changes that occur during the development of each type of cell; (3) functional differences between the types of cells as well as between cells of different ages; (4) additional studies on the biochemistry of intracellular digestion and nutrient distribution; (5) cell movement and the molecular bases of foreign material recognition and uptake; and (6) definition of the roles of such specialized cells as multinucleate macrocytes and serous cells.

References

Abolinš-Krogis, A. (1968). *In* "Studies in the Structure, Physiology and Ecology of Molluscs" (V. Tretter, Ed.), pp. 75–92. Academic Press, London and New York.

Acton, R. T. and Evans, E. (1968). *J. Bacteriol.* **95**, 1260–1266.

Alverdes, F. (1913). *Zeit. Wiss. Zool.* **105**, 598–633.

Andrew, W. (1965). "Comparative Hematology." Grune and Stratton, New York and London.

Bang, F. B. (1961). *Biol. Bull., Woods Hole* **121**, 57–68.

Beedham, G. E. (1965). *J. Zool.* **145**, 107–124.

Berthe, J. and Petitfrère, C. (1934a). *Ann. Physiol. Physiochim. Biol.* **10**, 975–977.

Berthe, J. and Petitfrère, C. (1934b). *Arch. Intern. Physiol.* **39**, 98–111.

Betances, L.-M. (1921). *Arch. Anat. microsc. Morph. exp.* **18**, 309–327.

Buchner, P. (1965). "Endosymbiosis of Animals with Plant Microorganisms." Interscience, New York.

Canegallo, M. A. (1924). *Riv. Biol.* p. 614.

Cheney, D. P. (1969). The morphology, morphogenesis and reactive responses of ^{3}H-thymidine labeled leucocytes in the manila clam, *Tapes semidecussata* (Reeve). Ph.D. thesis, University of Washington, Seattle, Washington.

Cheney, D. P. (1971). *Biol. Bull., Woods Hole* **140**, 353–368.

Cheng, T. C. (1966a). *Proc. Natl. Shellfish. Assoc.* **56**, 2.

Cheng, T. C. (1966b). *J. Invertebr. Pathol.* **56**, 111–122.

Cheng, T. C. (1967). *Adv. Mar. Biol.* **5**, 1–424.

Cheng, T. C. (1975). *Ann. N.Y. Acad. Sci.* **266**, 343–379.

Cheng, T. C. (1976a). *J. Invertebr. Pathol.* **27**, 125–128.

Cheng, T. C. (1976b). Proc. IXth Intl. Colloq. Invert. Pathol., pp. 190–194. Queens University Press, Kingston, Ontario, Canada.

Cheng, T. C. (1976c). *J. Invertebr. Pathol.* **27**, 263–268.

Cheng, T. C. (1977). *Comp. Pathobiol.* **3**, 21–30.

Cheng, T. C. and Auld, K. R. (1977). *J. Invertebr. Pathol.* **30**, 119–122.

Cheng, T. C. and Burton, R. W. (1965). *Chesapeake Sci.* **6**, 3–16.

Cheng, T. C. and Burton, R. W. (1966). *Parasitology* **56**, 111–122.

Cheng, T. C. and Cali, A. (1974). *Contemp. Top. Immunobiol.* **4**, 25–35.

Cheng, T. C. and Foley, D. A. (1975). *J. Invertebr. Pathol.* **26**, 341–351.

Cheng, T. C. and Galloway, P. C. (1970). *J. Invertebr. Pathol.* **15**, 177–192.

Cheng, T. C. and Howland, K. H. (1980). *J. Invertebr. Pathol.* In press.

Cheng, T. C. and Rifkin, E. (1970). *In* "Diseases of Fish and Shellfish" (S. F. Snieszko, Ed.) Am. Fisher. Soc. Symp. Vol. 5, 443–496.

Cheng, T. C. and Rodrick, G. E. (1974). *Biol. Bull., Woods Hole* **147**, 311–320.

Cheng, T. C. and Rodrick, G. E. (1975). *Comp. Biochem. Physiol.* **52B**, 443–447.

Cheng, T. C. and Rudo, B. M. (1976a). *J. Invertebr. Pathol.* **27**, 259–262.

Cheng, T. C. and Rudo, B. M. (1976b). *J. Invertebr. Pathol.* **27**, 137–139.

Cheng, T. C. and Sanders, B. G. (1962). *Proc. Penna. Acad. Sci.* **36**, 72–83.

Cheng, T. C. and Yoshino, T. P. (1976). *J. Invertebr. Pathol.* **27**, 243–245.

Cheng, T. C., Shuster, C. N. Jr. and Anderson, A. H. (1966). *Exp. Parasit.* **19**, 9–14.

Cheng, T. C., Thakur, A. S. and Rifkin, E. (1969). *In* "Proceedings of the Symposium on Mollusca," Part II. pp. 546–563. Bangalor, India.

Cheng, T. C., Cali, A. and Foley, D. A. (1974). *In* "Symbiosis in the Sea" (W. B. Vernberg, Ed.), pp. 61–91. University South Carolina Press, Columbia, South Carolina.

Cheng, T. C., Rodrick, G. E., Foley, D. A. and Koehler, S. A. (1975). *J. Invertebr. Pathol.* **25**, 261–265.
Cuénot, L. (1891). *Arch. Zool. Exper.* **9**, 19–54.
Dakin, W. J. (1909). *Liverpool Mar. Biol. Comm. Mem.* **17**, 1–144.
Dawson, A. B. (1932). *Biol. Bull., Woods Hole* **64**, 233–242.
De Bruyne, C. (1895). *Archs Biol.* **14**, 161–182.
Des Voigne, D. M. and Sparks, A. K. (1968). *J. Invertebr. Pathol.* **12**, 53–65.
Des Voigne, D. M. and Sparks, A. K. (1969). *J. Invertebr. Pathol.* **14**, 293–300.
Drew, G. H. (1910). *Quart. Jl. Microsc. Sci.* **54**, 605–621.
Dubois, R. (1901). *Compt. Rend. Heb. Séanc. Acad. Sci., Paris* **133**, 603–605.
Dubois, R. (1907). *Compt. Rend. Séanc. Soc. Biol.* **63**, 502–504.
Dunachie, J. F. (1963). *Trans. Roy. Soc. Edinb.* **65**, 383–411.
Dundee, D. S. (1953). *Trans. Am. Microsc. Soc.* **72**, 254–264.
Durham, H. E. (1891). *Quart. Jl. Microsc. Sci.* **33**, 81–121.
Durning, W. C. (1957). *J. Bone Joint. Surg.* **39A**, 377–393.
Eble, A. F. (1958). *Proc. Natl. Shellfish. Assoc.* **48**, 148–151.
Edelman, G. M. (1976). *Science* **192**, 218–226.
Esser, W. (1934). *Archs Biol.* **45**, 337–390.
Farley, C. A. (1968). *J. Protozool.* **15**, 585–599.
Feng, J. S. (1966). *J. Invertebr. Pathol.* **8**, 446–504.
Feng, S. Y. (1959). *Bull. N.J. Acad. Sci.* **4**, 17.
Feng, S. Y. (1965). *Biol. Bull., Woods Hole* **128**, 95–105.
Feng, S. Y. (1966). *J. Invertebr. Pathol.* **8**, 505–511.
Feng, S. Y. (1967). *Fed. Proc.* **26**, 1685–1692.
Feng, S. Y. and Feng, J. S. (1974). *J. Invertebr. Pathol.* **23**, 22–37.
Feng, S. Y. and Stauber, L. A. (1968). *J. Invertebr. Pathol.* **10**, 94–110.
Feng, S. Y., Feng, J. S., Burke, C. N. and Khairallah, L. H. (1971). *Z. Zellforsch. mikrosk. Anat.* **120**, 222–245.
Feng, S. Y., Feng, J. S. and Yamasu, T. (1977). *Comp. Pathobiol.* **3**, 31–67.
Fernau, W. (1914). *Zeit. Wiss. Zool.* **110**, 253–358.
Foley, D. A. and Cheng, T. C. (1972). *J. Invertebr. Pathol.* **19**, 383–394.
Foley, D. A. and Cheng, T. C. (1974). *Biol. Bull., Woods Hole* **146**, 343–356.
Foley, D. A. and Cheng, T. C. (1975). *J. Invertebr. Pathol.* **25**, 189–197.
Foley, D. A. and Cheng, T. C. (1977). *J. Invertebr. Pathol.* **29**, 321–325.
Franc, A. (1960). *In* "Traité de Zoologie, Anatomie, Systématique, Biologie" (P. P. Grassé, Ed.), Vol. 5, fasc. 2, pp. 1845–2133. Masson et Cie, Paris.
Galtsoff, P. S. (1964). *Fisher. Bull., Fish and Wildlife Serv.* **64**, 1–480.
George, W. C. and Ferguson, J. H. (1950). *J. Morph.* **86**, 315–324.
Giard, A. (1970). *Compt. Rend. Séanc. Soc. Biol.* **63**, 416–420.
Goetsch, W. and Scheuring, L. (1926). *Zeit. Morph. Okol. Tiere* **7**, 220–253.
Gomori, G. (1952). "Microscopic Histochemistry." University Chicago Press, Chicago, Illinois.
Goodrich, E. S. (1919). *Quart. Jl. Microsc. Sci.* **64**, 19–27.
Grassé, P. P. (1960). "Traité de Zoologie", Vol. 5, Masson et Cie, Paris.
Griesbach, H. (1891). *Arch. Mikroskop. Anat.* **37**, 22–99.
Grobben, C. (1888). *Arb. Zool. Inst. Univ. Wien* **7**, 355–444.
Haeckel, E. (1862). "Die Radiolarien." George Reiner, Berlin.
Haigler, S. A. (1964). A histochemical and cytological study of the "brown cells" found in the "auricular pericardial gland" and other tissues of the oyster, *Crassostrea virginica* (Gmelin). M.Sc. Thesis, University of Delaware, Newark, Delaware.

Harris, K. R. (1975). *Ann. N.Y. Acad. Sci.* **266**, 446–464.
Hazleton, B. J. and Isenberg, G. R. (1977). *Proc. Penna. Acad. Sci.* **51**, 54–56.
Hill, R. B. and Welsh, J. H. (1966). *In* "Physiology of Mollusca" (K. M. Wilbur and C. M. Yonge, Eds), Vol. II, pp. 125–174. Academic Press, New York and London.
Hopkins, A. E. (1934). *Science* **80**, 411–412.
Hopkins, A. E. (1936). *Biol. Bull., Woods Hole* **70**, 413–425.
Jameson, H. L. (1902). *Proc. Zool. Soc. Lond.* **1**, 140–166.
Jordon, H. E. (1952). "A Textbook of Histology." Appleton-Century-Crofts, New York.
Jullien, A. and Morin, G. (1931). *Comp. Rend. Hemdomad. Séanc. Mém. Soc. Biol. Filial.* **108**, 1242–1244.
Keber, G. A. F. (1851). "Beiträge sur Anatomie und Physiologie der Weichtiere." Gebr. Bornträger, Konigsberg.
Knoll, P. (1893). *Sitz.-Ber. Kais. Akad. Wiss. Wien 3 Abt.* **102**, 440–478.
Letellier, A. (1891). *Compt. Rend. Acad. Sci.* **112**, 56–58.
Lie, K. J., Heyneman, D. and Yau, P. (1975). *J. Parasitol.* **63**, 574–576.
Lille, R. D. (1954). "Histopathologic Technic and Practical Histochemistry." Blakiston, New York.
Mackin, J. G. (1951). *Bull. Mar. Sci. Gulf Caribb.* **1**, 72–87.
Manigault, P. (1939). *Ann. Inst. Oceanogr. Paris n.s.* **18**, 331–346.
Martin, A. (1970). *Proc. Natl. Shellfish. Assoc.* **61**, 5–6.
Martin, A. W. and Harrison, F. M. (1966). *In* "Physiology of Mollusca" (K. M. Wilbur and C. M. Yonge, Eds), Vol. II, pp. 353–386. Academic Press, New York and London.
McDade, J. E. and Tripp, M. R. (1967). *J. Invertebr. Pathol.* **9**, 531–535.
Michelson, E. H. (1961). *Am. J. Trop. Med. Hyg.* **10**, 423–427.
Mix, M. C. (1976). *Mar. Fisher. Rev.* **38**, 37–41.
Mix, M. C. and Sparks, A. K. (1971). *J. Invertebr. Pathol.* **17**, 172–177.
Mix, M. C. and Tomasovic, S. P. (1973). *J. Invertebr. Pathol.* **21**, 318–320.
Moore, C. A. and Eble, A. F. (1977). *Biol. Bull., Woods Hole* **152**, 105–119.
Moore, M. N. and Lowe, D. M. (1977). *J. Invertebr. Pathol.* **29**, 18–30.
Morton, B. (1969). *Proc. Malacol. Soc. London* **38**, 401–444.
Motley, H. L. (1933). *J. Morph.* **54**, 415–427.
Nakahara, H. and Bevelander, G. (1969). *Texas. Rep. Biol. Med.* **27**, 101–109.
Narain, A. S. (1968). *Gorakhpur Univ. Res. J.* **2**, 46.
Narain, A. S. (1969). Studies on the heart and blood of *Lamellidens corrianus* (Lea). Ph.D. Thesis, University of Gorakhpur, India.
Narain, A. S. (1972a). *J. Morph.* **137**, 63–70.
Narain, A. S. (1972b). *Experientia* **28**, 507.
Narain, A. S. (1973). *Malacol. Rev.* **6**, 1–12.
Naville, A. (1926). *Rev. Suisse Zool.* **33**, 251–289.
Ohuye, T. (1937). *Sci. Rep. Tôhoku Imp. Univ.*, ser. 4 **12**, 203–239.
Ohuye, T. (1938a). *Sci. Rep. Tôhoku Imp. Univ.*, ser. 4 **12**, 593–622.
Ohuye, T. (1938b). *Sci. Rep. Tôhoku Imp. Univ.*, ser. 4 **13**, 359–380.
Orton, J. H. (1923). *Fish Invest. London*, ser. 2 **6**, 3.
Owen, G. (1966). *In* "Physiology of Mollusca" (K. M. Wilbur and C. M. Yonge, Eds), Vol. II, pp. 53–96. Academic Press, New York and London.
Pan, C. T. (1958). *Bull. Mus. Comp. Zool. Harvard* **119**, 237–299.
Pauley, G. B. and Heaton, L. H. (1969). *J. Invertebr. Pathol.* **13**, 241–249.
Pauley, G. B. and Sparks, A. K. (1965). *J. Invertebr. Pathol.* **7**, 248–256.

Pearse, A. G. E. (1961). "Histochemistry: Theoretical and Applied." Little Brown, Boston.

Perrier, E. (1903). *Compt. Rend. Heb. Séanc. Acad. Sci., Paris* **137**, 682.

Prytherch, H. F. (1940) *J. Morph.* **66**, 39–64.

Purchon, R. D. (1968). "The Biology of the Mollusca." Pergamon Press, London.

Redfield, E. S. P. (1917). *J. exp. Zool.* **22**, 231–239.

Renwrantz, L. R. and Cheng, T. C. (1977). *J. Invertebr. Pathol.* **29**, 88–96.

Renwrantz, L. R., Yoshino, T. P., Cheng, T. C. and Auld, K. R. (1979). *Zool. Jahrb. Physiol.* **83**, 1–12.

Rifkin, E., Cheng, T. C. and Hohl, H. R. (1969). *J. Invertebr. Pathol.* **14**, 211–226.

Rodrick, G. E. and Cheng, T. C. (1974a). *J. Invertebr. Pathol.* **24**, 374–375.

Rodrick, G. E. and Cheng, T. C. (1974b). *J. Invertebr. Pathol.* **24**, 41–48.

Ruddell, C. L. (1971a). *J. Invertebr. Pathol.* **18**, 269–275.

Ruddell, C. L. (1971b). *Histochemie* **26**, 98–112.

Ruddell, C. L. (1971c). *J. Invertebr. Pathol.* **18**, 260–268.

Sato, T. (1931). *Zeit. Vergl. Physiol.* **14**, 763–783.

Schoenberg, D. A. and Cheng, T. C. (1980). *J. Invertebr. Pathol.* **36**, 141–143.

Singer, S. J. (1974). *Adv. Immunol.* **19**, 1–66.

Sioli, H. (1935). *Zool. Jahrb. Abt. Allgem. Zool. Physiol. Tiere* **54**, 507–534.

Sparks, A. K. (1972). "Invertebrate Pathology: Noncommunicable Diseases." Academic Press, New York and London.

Sparks, A. K. and Pauley, G. B. (1964). *J. Insect Pathol.* **6**, 78–101.

Stauber, L. A. (1950). *Biol. Bull., Woods Hole* **98**, 227–241.

Stauber, L. A. (1961). *Proc. Natl. Shellfish. Assoc.* **50**, 7–20.

Stein, J. E. and Mackin, J. G. (1955). *Texas J. Sci.* **7**, 422–429.

Takatsuki, S. I. (1934a). *Quart. Jl. Microsc. Sci.* **76**, 379–431.

Takatsuki, S. I. (1934b). *Sci. Rept. Tokyo Bunrika Daigaku, Sec. B* **2**, 55–62.

Tanaka, K. and Takasugi, T. (1964). *In* "Practical Shellfish Haematology" (K. Tanaka, Ed.), pp. 19–22. Ueda Bookstore, K. K., Japan.

Tanaka, K., Takasugi, T. and Maoka, H. (1961). *Bull. Jap. Soc. Sci. Fisher.* **27**, 365–371.

Tripp, M. R. (1958a). *Proc. Natl. Shellfish. Assoc.* **48**, 143–147.

Tripp, M. R. (1958b). *J. Parasitol.* **44** (Sect. 2), 35–36.

Tripp, M. R. (1960). *Biol. Bull., Woods Hole* **119**, 210–223.

Tripp, M. R., Bisignani, L. A. and Kenny, M. T. (1966). *J. Invertebr. Pathol.* **8**, 137–140.

Tsujii, T. (1960). *J. Fac. Fish. Pref. Univ. Mie* **5**, 1–70.

Turchini, J. (1923). *Archs Morph. Gén. exp.* **18**, 7–253.

von Brücke, E. T. (1925). *In* "Winterstein's Handbuch der vergleichenden Physiologie," Vol. 1, pp. 826–1110. Fischer, Jena.

von Buddenbrock, W. (1965). *In* "Vergleichende Physiologie," Vol. 6. Birkhäuser, Stuttgart.

Wagge, L. E. (1951). *Quart. Jl. Microsc. Sci.* **92**, 307–321.

Wagge, L. E. (1955). *Intl. Rev. Cytol.* **4**, 31–78.

White, K. M. (1942). *Proc. Malacol. Soc. London* **25**, 37–88.

Wilbur, K. M. (1964). *In* "Physiology of Mollusca" (K. M. Wilbur and C. M. Yonge, Eds), Vol. I, pp. 243–282. Academic Press, New York and London.

Yonge, C. M. (1923). *Brit. J. exp. Biol.* **1**, 15–63.

Yonge, C. M. (1926). *J. Mar. Biol. Assoc. U.K.* **14**, 295–388.

Yonge, C. M. (1936). *Sci. Rept. Great Barrier Reef Exped.* **1**, 283–321.

Yonge, C. M. (1937). *Biol. Rev. Cambridge Phil. Soc.* **12**, 87–115.
Yonge, C. M. (1946). *Nature* **157**, 729.
Yonge, C. M. and Nicholas, H. M. (1940). *Pap. Tortugas Lab.* **32**, 287–301.
Yoshino, T. P. and Cheng, T. C. (1976a). *Trans. Am. Microsc. Soc.* **5**, 215–220.
Yoshino, T. P. and Cheng, T. C. (1976b). *J. Invertebr. Pathol.* **27**, 367–370.
Yoshino, T. P. and Cheng, T. C. (1977). *J. Invertebr. Pathol.* **30**, 76–79.
Yoshino, T. P., Renwrantz, L. R. and Cheng, T. C. (1979). *J. exp. Zool.* **207**, 439–449.
Zacks, S. I. (1955). *Quart. Jl. Microsc. Sci.* **96**, 57–71.
Zacks, S. I. and Welsh, J. H. (1953). *Biol. Bull., Woods Hole* **105**, 200–211.

9. Cephalopods

R. R. COWDEN AND S. K. CURTIS

Department of Anatomy and Program in Biophysics, College of Medicine, East Tennessee State University, Johnson City, Tenn. 37614, U.S.A.

CONTENTS

I. Introduction

For a very long time, the haemopoietic tissues of cephalopod molluscs represented the only clearly identified blood-forming tissues known in any invertebrate. As such, they have received attention at various levels in the development of the technology of microscopy extending back into the last century. The most extensive study of cephalopod haemopoietic tissues using conventional methods of histology and cytology was published as a monograph in two sections by Bolognari (1949, 1951). He surveyed essentially all the cephalopod species that could be collected in Europe, and his description of cephalopod haemopoiesis, indeed the entire study, has to rank as a classic. While the authors of the present review disagree with Bolognari on some small matters of nomenclature, his observations are correct to the level that the technology of that time would allow. He also indicated that the blood-forming tissues are structurally similar in all cephalopods that have been studied. His bibliography refers the reader to the large number of earlier investigations of these tissues and circulating blood cells in cephalopods. More recently, Cowden (1968) and Cowden and Curtis (1973, 1974) have published the results of a light microscopic and cytochemical study of the blood-forming tissues of *Octopus vulgaris*; an electron microscopic survey of haemopoietic tissues and circulating blood cells in *Octopus briareus*; and an *in vitro* study of cells dissociated from these tissues. These respresent the most recent descriptions of the blood-forming cells and tissues of cephalopods.

II. Structure of the circulatory system

The anatomy of the circulatory system in cephalopod molluscs has been described in most standard textbooks of invertebrate zoology. An excellent account may be found, for instance, in the second edition of Borradaile and Potts (1951). Venous haemolymph, carried by the vena cava, the mantle veins, and the abdominal veins, empties into the paired, muscular gill-hearts (also called branchial-hearts) located at the base of the paired gills (Fig. 1). Each gill-heart propels the venous haemolymph through an afferent branchial vessel into the extensive capillary network of the gill where respiratory exchange occurs. The oxygenated haemolymph from each gill is carried through an efferent branchial vessel into one of the two auricles of the three-chambered heart and from thence into the unpaired median ventricle. Contraction of the ventricle distributes the haemolymph to the body through the anterior and posterior aortae (Fig. 1). The circulatory system in cephalopods is considered a closed system in that haemolymph is restricted to the blood vessels and capillaries. Cephalopods do not possess extensive haemocoels as are encountered in other molluscan groups.

The organization and ultrastructure of cephalopod blood vessels have received some recent attention from Barber and Graziadei (1965, 1967a,b). The major vessels are composed of an outer adventitial layer, a muscular middle layer and an inner endothelial layer. The muscular layer is thicker in arteries than it is in veins. The endothelial layer is considered incomplete since it does not cover the basal lamella in all locations. It is not unusual to find a

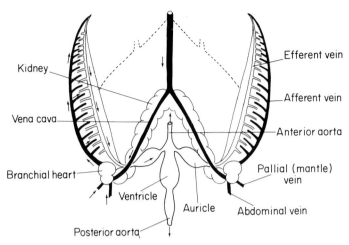

Fig. 1. Diagramme showing major vessels of the cephalopod circulatory system (modified from Steimpel, 1926 and Borradaile and Potts, 1951).

capillary consisting of only a basal lamella and some associated adventitial tissue. This unusual construction undoubtedly allows certain types of cells and particulates to pass directly in and out of the circulatory system.

The cells responsible for the production of haemocyanin, the characteristic copper-containing, oxygen-binding, respiratory protein of many molluscs and crustaceans, are found in the paired branchial glands, as indicated by Dilly and Messenger (1972). These paired glands, suspended from the mantle wall, are in close association with the dorsal surfaces of the gills, along their entire length. These ductless, highly vascularized glands were originally considered possible endocrine organs. Each gland is surrounded by a capsule composed of columnar epithelial cells on the outside and connective tissue internal to this investment. The organ appears to be filled homogeneously with polygonally-shaped gland cells, each of which typically possesses a vesicular nucleus and a prominent nucleolus. Haemocyanin has been identified in transmission electron micrographs both in the rough-surfaced endoplasmic reticulum and cytoplasmic vesicles of the cells. The haemocyanin apparently is released directly into the capillaries supplying the glands by a process of exocytosis. The haemocyanin-producing cells themselves do not enter the general circulation.

III. Structure and classification of haemocytes

Only a single type of blood cell can be found in the circulating haemolymph of mature cephalopods. This cell displays a mixture of the characteristics of vertebrate granulocytes (the presence of two classes of cytoplasmic granular inclusions) and of monocytes (the possession of an elongated, somewhat condensed, irregular or "U"-shaped nucleus). Romanovsky-type stains produce conspicuous staining of the spherical cytoplasmic inclusions by basic dye components and essentially "neutral" staining of the general cytoplasm, while the nuclei are also stained by the basic dye components (Fig. 2).

In "thick" sections of plastic-embedded material, the spherical inclusions are again the most conspicuous features of these cells (Fig. 3). The absolute amount of cytoplasm and the number and distribution of the specific granules are variable. At the ultrastructural level, a dominant class of granules can be seen; but while some of these appear to be homogeneously electron-dense, others display either variable density or have electron-dense cores (Fig. 4). In addition to these inclusions, the cells contain a few mitochondria; a Golgi apparatus that is substantially reduced in comparison with that observed in precursor cells (see below); and some rough-surfaced endoplasmic reticulum, as well as free ribosomes. Glycogen accumulations were not seen in mature haemocytes of *Octopus briareus*, the

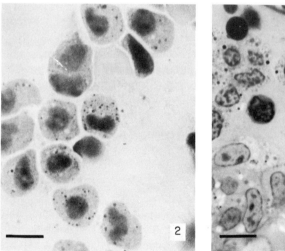
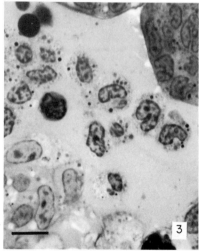

Fig. 2. Touch preparation of white body of *Octopus vulgaris* showing mature haemocytes stained with Azure A-Eosin B (Lillie) after Zenker-formol fixation. Line scale, 10 μm. × 1000.

Fig. 3. Plastic section of white body of *Octopus briareus* showing mature haemocytes stained with alkaline toluidine blue. × 1000.

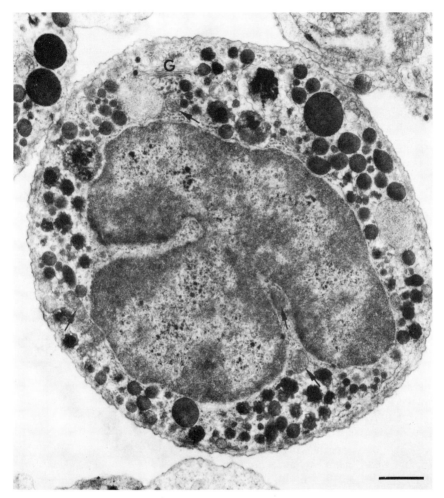

Fig. 4. Electron micrograph of a mature haemocyte of *Octopus briareus*. Note the predominance of spherical inclusion granules, some displaying irregular outlines and electron-dense cores. Cisternae of the rough-surfaced endoplasmic reticulum can be seen among the granules, and a small Golgi complex (G) is visible. Several mitochondria (arrows) are also present. The nucleus is irregular in shape; condensed chromatin is marginated against the nuclear membrane; and the extrachromosomal space contains small, very electron-dense granules. Line scale, 1 μm. × 12 000.

only species that has been subjected to recent ultrastructural examination (Cowden and Curtis, 1974). There are also no conspicuous microtubular or filamentous structures in these cells, but no special attempts were made to examine moving cells with extended pseudopodia, or cells in cellular clots. Further, the conditions used for fixation may not have been the most favourable for the preservation of such structures. The plasma membranes of these cells also display shallow, irregular indentations (Fig. 4), and conventional staining methods (lead citrate and uranyl acetate) did not reveal the presence of a conspicuously thickened carbohydrate coat (glycocalyx).

The nuclei of mature haemocytes are more compact than those encountered in most precursor stages. Most of the chromatin is condensed, especially along the inner surface of the nuclear membrane. The nucleoplasm also contains small, irregular and extremely electron-dense granules (Fig. 4). These granules are consistent features of these cells and those of the entire precursor series.

IV. Origin and formation of haemocytes

The blood-forming tissues of cephalopod molluscs are called "white bodies" (corpora bianca). They are located behind the eyes in the orbital pits (Fig. 12) and are composed of a cluster of somewhat flattened multi-lobed (usually three) structures extending about a third of the perimeter of the orbit. They appear as white, almost fatty tissue. There are no studies of the embryonic development of white bodies, and there are no accounts of the early growth of blood-forming tissues in the newly hatched cephalopod.

The flattened lobes of the white bodies are surrounded by a layer of fibrous connective tissue, and the lobes are traversed by major vessel systems (arteries and veins which arborize into the tissue). Throughout the lobular tissue there are zones containing cells in various stages of development, as well as sinusoids into which the maturing and matured haemocytes enter, presumably on their way into the circulating haemolymph.

The precursor classes are associated chiefly with vascular arborizations; and in preparations stained for the demonstration of RNA, the higher concentration of the RNA characteristic of the most primitive precursor and the next two or three levels of differentiation is especially useful for visualizing the general organization of the tissue (Fig. 5). Additional information about the general organization of the tissue can be obtained from sections of tissue fixed in Zenker-formol and stained by Lillie's azure A-eosin B Romanovsky method. Even though staining for RNA is less pronounced after fixation in formalin, there is superior preservation of cytoplasmic structures and dimensional relationships (Fig. 6).

While Bolognari (1949, 1951) reported that the white bodies are inter-laced with fibrous connective tissue and reticular fibres, our observations made on both *Octopus vulgaris* and *O. briareus* using Bouin-fixed material and Masson's trichrome connective tissue stain, or a modification of Foote's silver stain for reticular fibres, indicated that fibrous connective tissue is associated mainly with the capsule and the vascular tracts. At the ultra-structural level, only occasional collagen fibrils or bundles of fibrils were seen. Nevertheless, in spite of a virtual absence of intralobular connective tissue, the tissue is quite compact and resists facile mechanical dissociation.

From a combination of ultrastructural studies and the examination of "thick" plastic sections, some of the non-haemopoietic cells of the white bodies can be identified. In addition to the cells which construct the blood

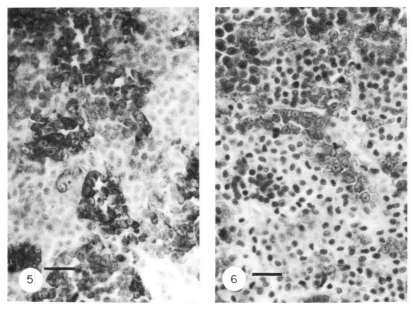

Fig. 5. Paraffin section of white body of *Octopus briareus* showing high concen-tration of RNA in precursor cells (haemocytoblasts, primarily and secondary leuco-blasts) around vascular arborizations. Line scale, 20 μm. × 400.

Fig. 6. Paraffin section of *Octopus vulgaris* white body stained with Azure A-Eosin B after Zenker-formol fixation. There is superior preservation to that in Fig. 5. Note large nuclei of more primitive precursor cells and smaller, denser nuclei of the more mature haemocytes. × 400.

vessels, the main non-haemopoietic cells that can be found are fibroblasts. These cells are not abundant.

The most primitive cell in the series which progressively differentiate into leucocytes has been designated variously as a "haemocytoblast", "primary leucoblast" or "reticulum cell". The authors of the present chapter prefer the more generalized term "haemocytoblast", and cells of this kind will be referred to as such throughout the remainder of this review. Haemocytoblasts differentiate into primary and secondary leucoblasts which in turn develop into mature leucocytes (haemocytes) via transitional forms.

A. Haemocytoblasts

Haemocytoblasts contain a very large nucleus which generally possesses a single, large basophilic nucleolus. Such cells may also be recognized in light microscopic preparations because of their very large size (Fig. 6). In preparations stained for DNA with the Feulgen reaction or by gallocyanin-chromalum after pretreatment with RNase, again they are readily recognized by their large size and extensive nucleolus-associated heterochromatin. These cells were never seen in any stage of mitosis.

In all haemocytoblasts examined with the electron microscope (Fig. 7), the nucleus is extremely large, and only a small amount of chromatin appears to be condensed against the inner margins of the nuclear envelope. A substantial amount of chromatin is associated with the nucleoli; and these structures often appear somewhat asymmetrical in shape and lack a clear organization into a typical *pars fibrosa* and *pars granulosa*. The extrachromatin space contains the irregular and variably sized, extremely electron-dense nuclear granules noted earlier (Fig. 7).

The plasma membranes of these cells are generally in close apposition with those of adjacent cells; and the surfaces of the cells are covered with a thin coating of an electron-dense material resembling glycoprotein. A particularly interesting and unusual feature of the cells are the well-developed intercellular bridges (Fig. 7), similar to those that have been described in developing germ cells (Fawcett et al., 1959; Ruby et al., 1969, 1970a,b; Skalko et al., 1972). In favourable sections of *Octopus* haemocytoblasts, the walls of the intercellular bridges appear to be constructed of a series of annulae spaced at a fixed periodicity (Fig. 7). The cytoplasm of the haemocytoblasts contains a moderate amount of rough-surfaced endoplasmic reticulum, mostly concentrated around the nucleus and toward the periphery of the cell. The Golgi apparatus consists of stacks of lamellae with terminal vesicles containing extremely electron-dense material. Slightly

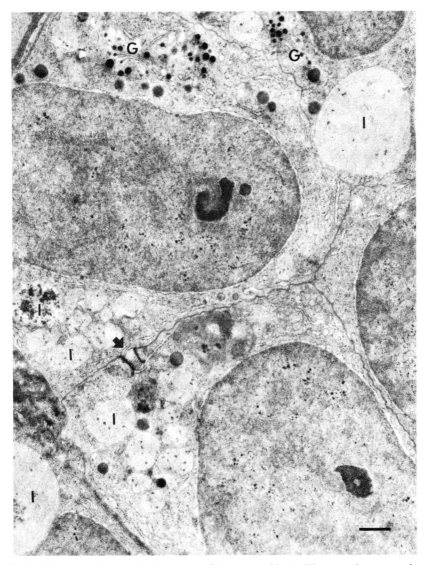

Fig. 7. Electron micrograph of a group of haemocytoblasts. The cytoplasm contains flattened cisternae of the rough-surfaced endoplasmic reticulum; Golgi elements (G), including electron-dense vesicles; and large spherical inclusions (I) in which electron-dense material is deposited in some instances. An intercellular bridge (arrow) can be seen between two of the cells. This bridge apears to be constructed of stacked annulae superimposed on a denser external matrix. The rounded nuclei of the cells display a predominance of loosely organized chromatin with little packing along the inner surfaces of the nuclear membranes. Nucleoli, visible in two of the cells, are irregular in shape and lack a distinct organization into a *pars fibrosa* and *pars amorpha*. The extrachromosomal nucleoplasm contains electron-dense granules. Line scale, 1 μm. × 8000.

larger, very dense spherical inclusions are also seen frequently in the vicinity of the Golgi apparatus. In addition, the most commonly observed inclusions of haemocytoblasts are larger and spherical granules which display very little electron-density (Fig. 7). These inclusions often contain either a finely coiled, tubular substructure, or, less frequently, variable and irregular accumulations of electron-dense material. Although mitochondria are also present, as well as some free ribosomes, the cytoplasm is remarkably free of electron-dense material. This probably accounts for the general paleness of these cells in "thick" plastic sections.

B. Primary and secondary leucoblasts

Bolognari (1951) recognized three stages of leucoblast development beyond the haemocytoblast, and there is some support for this classification scheme at both the light and electron microscopic levels. However, cells of these types are intergrades between extremes and possess no particular set of characteristics, other than subtle differences in size or nuclear-cytoplasmic ratio, that would permit the identification of clear intermediate stages. Thus, only primary and secondary leucoblasts have been recognized in the present review, with the understanding that criteria might evolve later that would allow further enlargement of this precursor category.

Primary leucoblasts closely resemble haemocytoblasts, except that they are smaller and appear to be associated in clusters. They are found much more frequently than haemocytoblasts. Like haemocytoblasts, they lack conspicuously clumped and marginated chromatin; and they posess nucleolus-associated heterochromatin and irregularly shaped nucleoli which lack a conventional concentric arrangement of granular and fibrous elements (Fig. 8). As in all other stages, the electron-dense granules are present in the nucleoplasm. Although primary leucoblasts contain larger concentrations of rough-surfaced endoplasmic reticulum than haemocytoblasts the kind and classes of cytoplasmic inclusions are similar. The large spherical inclusions, however, often contain a precipitate associated with the coiled tubular substructure, and a few contain a very dense central core (Fig. 8). The Golgi apparatus is often highly developed and intercellular bridges are always observed (Fig. 8). It seems obvious that the bridges serve as potential avenues for the flow of molecular information between developing cells, and they also probably contribute to the difficulty experienced in attempts to dissociate the tissue mechanically. From the inspection of preparations pretreated with DNase and stained with azure B, it is clear that primary leucoblasts exhibit the highest overall amounts of cytoplasmic RNA. As in the case of haemocytoblasts, the cells were not seen in mitosis.

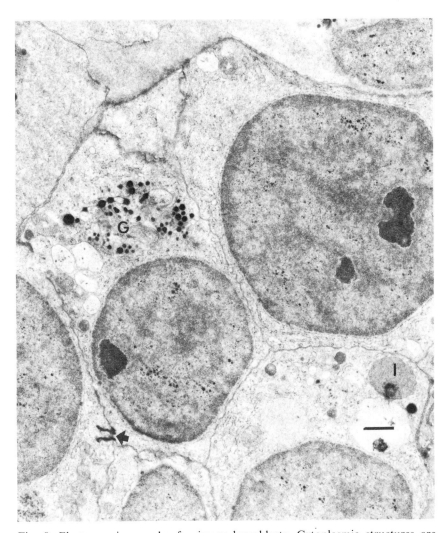

Fig. 8. Electron micrograph of primary leucoblasts. Cytoplasmic structures are similar to those of haemocytoblasts, although there is some increase in rough-surfaced endoplasmic reticulum; and vesicular components of the Golgi complex (G) are particularly prominent in one of the cells. Large spherical inclusions (I) with electron-dense cores are present. Two of the cells are joined by an intercellular bridge (arrow). The chromatin is more condensed than that of haemocytoblasts, especially in the vicinity of nucleoli. The nucleoli are irregular in shape and lack a clear separation of amorphous and fibrillar components. Extrachromosomal electron-dense granules are present in the nucleoplasm. Line scale, 1 μm. × 8000.

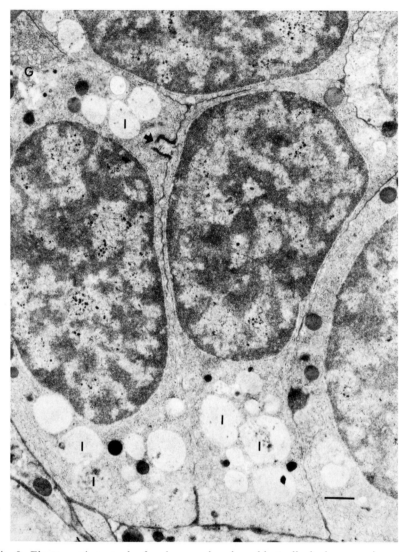

Fig. 9. Electron micrograph of early secondary leucoblasts displaying some increase in electron-dense cell-surface material as well as greater accumulations of large inclusions (I) containing precipitated material in some instances. A second class of smaller electron-dense spherical granules can also be seen. Golgi elements (G) as well as cisternae of the rough-surfaced endoplasmic reticulum are visible; and two of the cells are joined by an intercellular bridge (arrow). The nuclei possess a greater amount of condensed chromatin than at earlier stages, especially along the inner surfaces of the nuclear membranes. Electron-dense extrachromosomal granules are present, as in earlier stages. Line scale, 1 μm. × 8000.

Secondary leucoblasts are smaller cells with relatively less cytoplasm (Fig. 9). At this stage, the main amplification in cell numbers occurs by mitosis. In electron micrographs, the surfaces of these cells are covered by a somewhat thickened layer of electron-dense material as compared to earlier stages; and the characteristic intercellular bridges are present, again presenting the impression that they are constructed of stacked annulae. The chromatin is obviously more condensed than that of earlier stages, not only against the nuclear envelope, but also throughout the entire nucleus in a rather regular branched network (Fig. 9). Nucleoli are present, but they are less prominent than those observed in earlier stages. Secondary leucoblasts also contain a larger amount of rough-surfaced endoplasmic reticulum, a few mitochondria, elaborate Golgi complexes, and an overall cytoplasmic electron-density greater than that of earlier stages. In addition, another class of cytoplasmic inclusions is present, namely, smaller, electron-dense spherical bodies (Fig. 9). While inclusions of this type are encountered at all stages, they begin to achieve equivalence with or outnumber the larger, less electron-dense spherical inclusions during the secondary leucoblast stage (Fig. 10).

C. Transitional cells to leucocytes (haemocytes)

As cellular differentiation progresses, the amount of cytoplasm increases and the nuclei become more compact and condensed. These changes are accompanied by a progressive reduction in cytoplasmic RNA levels, the loss of nucleoli, and an increase in cell-specific products. (See Campbell and Gledhill, 1973, for a similar description for the differentiation of erythrocytes.) Thus amplification in cell numbers occurs in the pre-terminal cells, the secondary leucoblasts. The amount of cytoplasm enlarges in post-mitotic cells as they mature into leucocytes. Multiple and very complex Golgi bodies are seen frequently in transitional cells; and the chromatin in the now enlarging nuclei becomes less condensed. The shape of the nucleus also becomes irregular and it may elongate and flex into a "U"-shaped form (Fig. 11). During the transformation of secondary leucoblasts into leucocytes the nucleoli are lost

The transitional cells also contain cytoplasmic organelles required to support additional synthesis, including larger quantities of rough-surfaced endoplasmic reticulum, more mitochondria and multiple Golgi complexes. In mature cells observed in the sinusoids, the plasma membrane displays a smoothly undulating surface, but it lacks distinct microvilli and pinocytotic vesicles. The lack of a thick, conspicuous layer of microtubular material beneath the plasma membrane clearly distinguishes these cells from vertebrate

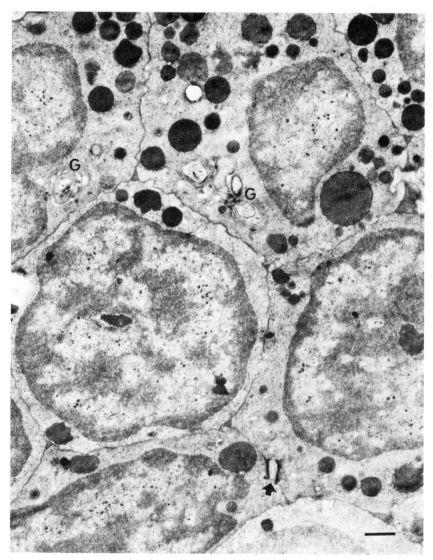

Fig. 10. Electron micrograph of more mature secondary leucoblasts displaying an increased number of small spherical electron-dense granules, some with slightly irregular outlines. Golgi complexes (G) are visible in two of the cells; and an intercellular bridge (arrow) is especially conspicuous. Nuclei display condensed chromatin along the inner surfaces of the nuclear membranes. Nucleoli are compact and irregular in shape, lacking a clear separation of the amorphous and fibrillar components. Dense extrachromosomal granules are also present. Line scale, 1 μm. × 8000.

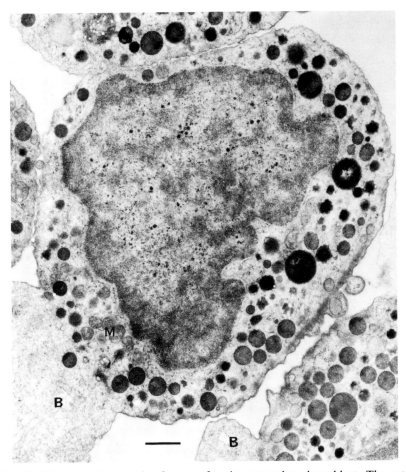

Fig. 11. Electron micrograph of a transforming secondary leucoblast. The cytoplasm contains large numbers of small, electron-dense spherical inclusions in a variety of configurations: smooth, less dense spheres; larger, very dense spheres with internal irregularities; and irregularly shaped granules with electron-dense cores. Small mitochondria (M) are often found in clusters. Note cytoplasmic blebs (B). The nucleus displays condensed, marginated chromatin and extrachromosomal dense granules. While small nucleoli are found occasionally in transforming cells, they are usually absent, as in this case. Note the irregular shape of the nucleus. Line scale, 1 μm. ×9500.

neutrophils. Cytoplasmic blebs of substantial size may arise from these cells (Fig. 11). These blebs usually do not contain conventional organelles, but they may contain microtubular components. The transformation of secondary leucoblasts into mature cells is accompanied by an increase in the number of small, spherical, electron-dense inclusions in the cytoplasm and by a tendency of some of these inclusions to become more irregular in shape (Fig. 11). The irregular inclusions often contain central zones of very great electron-density. Although Cowden (1968) reported the presence of amylase-sensitive, periodic acid-Schiff (PAS)-stainable material in *Octopus vulgaris* haemocytes (presumably glycogen), no glycogen has been observed in the haemocytes of *Octopus briareus* at either the light or electron microscopic levels. This may have been due to some seasonal or environmental variation in these animals, and a final conclusion concerning the amount of glycogen stored by the haemocytes must await further comparative information.

D. Summary

The white bodies, which lie behind the eyes in the orbital pits of cephalopod molluscs, produce a single type of granulated haemocyte which is released into the general circulation. Within the white body, the large haemocytoblast is the most primitive cell in the developmental series. In common with the most primitive generative cells in vertebrates, these cells and their immediate descendants, the primary leucoblasts, are seldom encountered in mitosis (never in the authors' material). All of the precursor cells, except the free transitional cells and the mature leucocytes, are connected by intercellular bridges. In contrast to haemocytoblasts and primary leucoblasts, secondary leucoblasts proliferate actively and form the main cellular population that amplifies the number of cells available for transition into definitive leucocytes. This transition involves an enlargement of the cytoplasm; an accumulation of additional small, cell-specific, spherical granules; some enlargement, elongation and folding of the nuclei; and a loss of nucleoli.

V. Functions of haemocytes

It has been difficult to ascribe any particular functions to mature cephalopod haemocytes beyond the formation of cellular clots. The *in vitro* studies of Cowden and Curtis (1973) indicated that these cells rapidly aggregate into spherical clots, the integrity of which is sufficient to prevent the penetration of the fluorescent dye acridine orange into the clots more than two or three cell layers. Since these clots have not yet been studied by electron microscopy, the anatomical basis for such slow penetration of acridine orange is not known.

There is, however, some considerable evidence for encapsulation and inflammation reactions in cephalopods. Jullien (1928) indicated that "granulocytes" of cephalopods accumulated around implanted sterile sutures and encapsulated the foreign bodies. There seems to be little doubt that the cells forming the capsule were haemocytes, although it is puzzling why the granules of the haemocytes were stained with an acidic dye (eosin). Conceivably, the preparations may have been stained by a method that resulted in the generalized binding of acidic rather than basic dyes to most cytoplasmic structures. It should also be noted that in touch preparations, smears and paraffin sections, only *some* of the inclusions, mainly the larger ones, are stained by the basic dye components of Romanovsky mixtures; however, virtually all cellular structures are stained with basic dyes at an alkaline pH in "thick" plastic sections. While in Jullien's (1928) work the granulocytes (= haemocytes) formed the primary capsule, connective tissue cells contributed collagenous secondary capsule material. Eventually, the collagenous capsule was eroded and the sutures were expelled from the tissue. In later experiments, a series of cephalopods, including *Sepia* sp. and *Eledone* sp., were injected with tar, a putative carcinogen. In all cases, the animals treated these injections as simple inflammatory challenges. The reactions and sequence of events were identical, but the reaction was more rapid in *Sepia* sp. than in *Eledone* sp. Jacquemain *et al.* (1947) inserted crystals of 1,2,5,6-dibenzanthracine in Vaseline into the mantle tissue of *Sepia* and obtained a reaction that appeared to be similar to that observed in oysters when their tissues were injected with terpentine (Sparks, 1972). The tissue developed a rapid oedema, accompanied by substantial histolysis. The entire area was sloughed and eventual healing occurred if the tissue loss were not too great. In the main, these French investigators demonstrated that the responses of cephalopods to inflammatory insults fit into patterns that were subsequently established for molluscs in general (Sparks, 1972). They were not able to induce neoplasia in cephalopods by these various experimental manipulations.

Regarding phagocytosis by the circulating haemocytes, although in *Eledone cirrosa* the blood cells have been shown to take up human erythrocytes in the presence of a serum factor, clearance of foreign particles injected *in vivo* appears to be mediated in cephalopods by other mechanisms. Thus both Stuart (1968) working on *E. cirrosa* and Bayne (1973a) on *Octopus dofleini* have shown the rapid removal from the blood of injected colloidal carbon or bacteria, without apparently a significant phagocytic involvement of the circulating blood cells. In the case of the colloidal carbon, this response was traced to phagocytes located in the gills, posterior salivary gland, white body and anterior salivary gland (Figs 12, 13) (Stuart, 1970; Bayne, 1973a). Stuart (1970) postulated that *E. cirrosa* therefore had a primitive reticuloendothelial system. It was not determined whether the phagocytic cells in these tissues were true fixed phagocytes or merely particle-laden phagocytes which had settled out of circulation at specific sites in the body. Bayne (1973a) also showed that prior exposure to bacteria did not enhance the clearance rates so that probably no adaptive immunity or immunological memory exist.

It may well be possible to elucidate other functions of the haemocytes by the use of tissue culture techniques. Necco and Martin (1963) utilized tissue culture medium NTC-199 containing a large quantity of additional sodium chloride (26·833 g/litre) to establish *in vitro* cultures of cells dissociated from

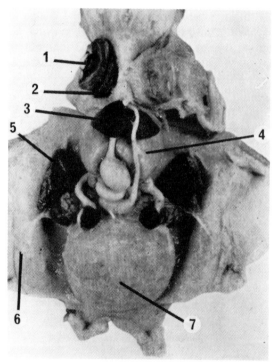

Fig. 12. Dissection of *Eledone cirrosa* two days after intravenous injection of high dose of carbon. The salivary gland, gills and white body are deeply pigmented. 1, Eye; 2, white body; 3, posterior salivary gland; 4, liver; 5, gill (ctenidium); 6, mantle; 7, ovary. (From Stuart, 1970).

the white bodies of *Octopus vulgaris*. Mitotic figures were noted among secondary leucoblasts, and incorporation of tritiated thymidine into the same cells was demonstrated. Survival was noted up to nine days, but no special attempts made to examine the short-term fate of the cells that divided (or of those which synthesized DNA). It would appear that this medium will keep the cells alive and in reasonable condition for a specific period, but it lacks critical factors required to sustain proliferation and development. A form of organ culture was employed by Durchon and Richard (1967) and Gomot and Guyard (1968) using a medium developed by Guyard (1967) which allowed the demonstration that the optic glands of *Sepia officianalis* when removed from control of the central nervous system are induced to produce secretory products. By placing ovarian fragments in the presence of stimulated optic glands, it was further demonstrated that these products stimulated mitotic division among ovarian follicle cells and vitellogenesis in oocytes which were greater than 0·3 mm in diameter. This endocrinological study points to the general usefulness of *in vitro* systems as well as the need to consider humoral factors in studying processes of the kinds that might be expected in haemocyte development. In studies of

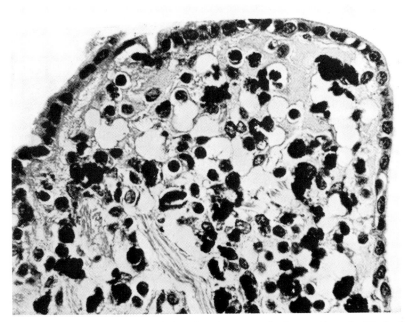

Fig. 13. Section of *E. cirrosa* gill showing the stromal phagocytic cells filled with carbon a few hours after intravenous injection. (From Stuart, 1970). × 800.

mammalian blood cell development, Bradley *et al.* (1969) have demonstrated the requirement for Colony Stimulating Factor (CSF) for *in vitro* proliferation of monocytoblasts and myeloblasts, and it would not be unreasonable to expect that some humoral factors might be required for the proliferation and maturation of cephalopod haemocytes *in vitro*.

VI. Comparison with other molluscs and concluding remarks

It would be imprudent to consider experimental work on immunity, cellular defence and haemopoiesis in cephalopod molluscs without considering work that has been published on these processes in other molluscan groups. Since some of these molluscan species have economic importance or serve as intermediate hosts for parasites of man or domestic animals, there has been a great deal more experimental work dedicated to lamellibranchs and gastropods than to cephalopods.

Cheng (1969) examined the electrophoretic patterns of haemolymph proteins of the large marine prosobranch, *Heliosome duryi normale*, which had been singly or repeatedly injected with bacteria and concluded that no new protein was being produced in response to this antigenic challenge. His studies confirmed the findings of earlier workers on other molluscs which indicated that while these animals may produce natural agglutinins, they do not produce immunoglobulins, or substances that behave as antibodies. Bayne (1973b, 1974) demonstrated that bacteria injected into *Helix pomatia* are removed within about two hours. When these animals are re-injected after a reasonable interval, the kinetics of removal are identical. He found essentially similar results with *O. dofleini* (see above), but while the main accumulation of phagocytosed bacteria was found in fixed cells of the gills of *O. dofleini*, they were sequestered chiefly in the digestive glands and other organs of *H. pomatia*. Curtis and Cowden (1978) reported that when the large slug, *Limax maximus*, is injected with fluorescent-labelled mammalian proteins at intervals of two hours, the proteins are phagocytosed into connective tissue "pore cells." In *L. maximus*, large numbers of these cells are present in the adventitia of blood vessels and within the pulmonary sac, but "pore cells" also occur in reasonable numbers throughout the animals in the loose connective tissue. The two xenogenic proteins (which are conjugated with contrasting fluorescent labels) were incorporated into phagolysosomes and destroyed in about two weeks. All reacting cells appeared to contain both kinds of proteins, so that neither dose appeared to saturate the system. Crichton and Lafferty (1976) demonstrated that the system of fixed macrophage in members of the most primitive class of molluscs, the *Amphineura*, is capable of discriminating among very closely related xenogenic proteins. They injected iodinated "self" haemocyanin and haemocyanin from three other species of chitons—each taxonomically more remote from the species being tested—and from the Australian crayfish. The kinetics of removal were determined in each instance. "Self" protein was tolerated indefinitely, while crayfish haemocyanin was removed very rapidly—within about two hours. Haemocyanin from the most taxonomically remote chiton was removed at a relatively rapid rate, while that from a near relative was removed quite slowly. Haemocyanin from a taxonomically intermediate chiton was removed at a rate that was intermediate between the other two.

Some generalizations appear to be emerging that are probably valid for the Phylum Mollusca as a whole. Molluscs do not make antibodies or antibody-like substances, although natural agglutinins are produced frequently. In cellular defence they do not appear to exhibit immunological memory. On the other hand, they possess a capacity for the removal of massive doses of bacteria or foreign materials and a very delicate capacity to discriminate among various types of xenogenic macromolecules. This latter position is put forward cautiously, because only large proteins have been considered thus far. While haemocytes may not participate in phagocytosis to the same extent in all molluscan species, in cephalopods they do not appear to have a major role in immediate phagocytosis. This does not rule out the possibility that after encapsulating foreign substances or particulates they might subsequently carry out phagocytosis.

The aggregation of molluscan haemocytes to form "cellular clots" has been studied most extensively in limpets by Davies and Partridge (1972), Partridge and Davies (1974), Jones and Partridge (1974) and Jones et al. (1976). The haemocytes (= granulocytes) of limpets might, or might not, be homologous to the haemocytes of cephalopods. Limpet haemocytes are actively stimulated to aggregate upon withdrawal from the circulating haemolymph. While aggregation will occur at $0°C$, it is accelerated at $20°C$. The ability of the cells to spread out on surfaces is affected by the presence of calcium and magnesium ions, but these ions do not affect "spike" formation. Colchicine causes the cells to lose bipolar form, and the entire margin is then supported by a lamella which normally forms the leading edge of the cell during "spike" formation. On the basis of morphometric studies of the aggregating cells at the ultrastructural level, these investigators have put forward a theory that lamella/cell body, lamella/lamella and body/cell body contacts have a quantitative role in adhesion which is in turn related to the underlying cytoskeletal elements.

It would be interesting to determine the response in more detail of cephalopods to xenoplastic transplants, the longer-term alterations of haemocytes in association with foreign substances, and the extent to which the haemocytes of cephalopods correspond to the granulocytes of other molluscan groups. It would also be useful to know how the experimental manipulation of cells in the white bodies or the circulating haemocytes would affect the precursor and circulating haemocyte populations. Related to this, it would seem probable that cephalopod tumours, including haemopoietic varieties, exist in nature. In a survey of tumours in lamellibranchs largely obtained from fisheries pathologists, Pauley (1969) noted the occurrence of haemopoietic tumours in oysters from various locations, including the Chesapeake Bay, the Pacific Coast of the United States and Australia. It might be possible to induce transplantable haemopoietic neoplasms in cephalopods, and these in turn might offer extremely useful experimental tools for future research. Balls and Rubin (1976) have proposed that the molluscs, because of their high level of tissue organization, might offer the most profitable group among the invertebrates in which to study neoplasia and cellular responses to neoplasia.

References

Balls, M. A. and Rubin, L. N. (1976). In "Comparative Immunobiology" (J. J. Marchalonis, Ed.), pp. 167–208. Halstead Press, New York.

Barber, V. C. and Graziadei, P. (1965). Z. Zellforsch. mikrosk. Anat. **66**, 765–781.

Barber, V. C. and Graziadei, P. (1967a). Z. Zellforsch. mikrosk. Anat. **77**, 147–161.

Barber, V. C. and Graziadei, P. (1967b). Z. Zellforsch. mikrosk. Anat. **77**, 162–174.

Bayne, C. J. (1973a). Malocological Rev. **6**, 13–17.

Bayne, C. J. (1973b). J. comp. Physiol. **86**, 17–25.

Bayne, C. J. (1974). In "Contemporary Topics in Immunobiology" (E. L. Cooper, Ed.), Vol. 4, pp. 37–45. Plenum Press, New York.

Bolognari, A. (1949). Arch. Zool. Ital. **34**, 79–97.

Bolognari, A. (1951). Arch. Zool. Ital. **36**, 253–285.

Borradaile, L. A. and Potts, F. A. (1951). "The Invertebrata", 2nd edn. Cambridge University Press, Cambridge.

Bradley, T. R., Metcalf, D., Sumner, M. and Stanley, R. (1969). "Hemic Cells", P. Farnes, In Vitro **4**, 22–35. Williams and Wilkins, Baltimore.

Campbell, C. L. and Gledhill, B. L. (1973). Chromosoma **41**, 385–394.

Cheng, T. C. (1969). J. Invert. Path. **14**, 60–81.

Cowden, R. R. (1968). J. Invert. Path. **19**, 113–119.

Cowden, R. R. and Curtis, S. K. (1973). Exp. Mol. Path. **19**, 178–185.

Cowden, R. R. and Curtis, S. K. (1974). In "Contemporary Topics in Immunobiology" (E. L. Cooper, Ed.), Vol. 4, pp. 77–90. Plenum Press, New York.

Crichton, R. and Lafferty, K. J. (1975). In "Immunologic Phylogeny" (W. H. Hildeman and A. A. Benedict, Eds), pp. 89–98. Plenum Press, New York.

Curtis, S. K. and Cowden, R. R. (1978). Dev. Comp. Immunol. **2**, 727–733.

Davies, P. S. and Partridge, T. (1972). J. Cell Sci. **11**, 757–769.

Dilly, P. N. and Messenger, J. B. (1972). Z. Zellforsch. mikrosk. Anat. **132**, 193–201.

Durchon, A. and Richard, M. (1967). *C.R. Acad. Sci., Ser. D* **204**, 1497–1500.

Fawcett, D. W., Ito, S. and Slautterback, D. L. (1959). *J. Biophys. Biochem. Cytol.* **5**, 453–460.

Gomot, L. and Guyard, A. (1968). *Proc. Intern. Colloq. Invertebrate Tissue Culture*, 2nd Conference (Abst.).

Guyard, A. (1967). *C.R. Acad. Sci., Ser. D* **265**, 147–149.

Jacquemain, R., Jullien, A. and Noel, R. (1947). *C.R. Acad. Sci. Ser. D* **225**, 441—443.

Jones, G. E. and Partridge, T. (1974). *J. Cell Sci.* **16**, 385–399.

Jones, G. E., Gillett, R. and Partridge, T. (1976). *J. Cell Sci.* **22**, 21–33.

Jullien, A. (1928). *C.R. Acad. Sci., Ser. D* **186**, 256.

Necco, A. and Martin, R. (1963). *Expl. Cell Res.* **30**, 588–590.

Partridge, T. and Davies, P. S. (1974). *J. Cell Sci.* **14**, 319–330.

Pauley, G. B. (1969). *In* "Neoplasms and Related Disorders of Invertebrates and Lower Vertebrates". *Nat. Canc. Inst. Monogr.* **31**, 509–539.

Ruby, J. R., Dyer, R. F. and Skalko, R. G. (1969). *J. Morph.* **127**, 307–340.

Ruby, J. R., Dyer, R. F., Skalko, R. G. and Volpe, E. P. (1970a). *Anat. Res.* **167**, 1–10.

Ruby, J. R., Dyer, R. F., Gasser, R. S. and Skalko, R. G. (1970b). *Z. Zellforsch. mikrosk. Anat.* **105**, 252–258.

Skalko, R. G., Kerrigan, J. M., Ruby, J. R. and Dyer, R. F. (1972). *Z. Zellforsch. mikrosk. Anat.* **128**, 31–41.

Sparks, A. K. (1972). "Invertebrate Pathology: Noncommunicable Diseases". Academic Press, New York.

Steimpel, W. (1926). "Zoologie im Grundriss". Borntraeger, Berlin.

Stuart, A. E. (1968). *J. Path. Bact.* **96**, 401–412.

Stuart, A. E. (1970). "The Reticulo-endothelial System". Livingstone, Edinburgh and London.

Taxonomic Index

li

T

U

Subject Index

A

Acanthocephala, 11
Accessory heart, 240, 424
Acid β-N-glucosaminidase, 400
Acid cell, 143, 144, 154
Acid hydrolase, 269, 270, 285
Acid phosphatase, 268, 275, 432, 433, 436, 458, 460, 630, 631
  cellular and serum components, 293
  chloragosomes of *Lumbricus*, 106
  granules, in acidophils of *Crassostrea*, 265
  granules, in gastropod amoebocytes, 205
  granules, in gastropod granulocytes, 202
  granules, in *Lumbricus* neutrophils, 94, 135
  granulocytes and macrophages, different levels in, 262
  granulocytes of Sipuncula, 176
  infrequent in *Lumbricus* granulocytes, 101
  low-to-moderate frequency in *Lumbricus* acidophils, 101
  lysosomes of *Lumbricus* basophils, 89, 135
  peritoneal coelomocytes of Polychaeta, 62
  phagocytosing amoebocytes, 216
  phagocytosis, activity stimulated by, 216
  polychaete amoebocytes, 64
  titre higher in haemocytes than plasma, 400
Acidophil, 265
  cytochemistry, in *Lumbricus*, 93, 100, 101
  cytology, in *Lumbricus*, 95
  frequency of, in Bivalvia, 254
  granular amoebocyte type, in oligochaetes, 81, 84, 118, 123, 135, 136
  Lumbricus terrestris, Types I, II, 86, 87, 93, 95, 98–101

  phagocytosis, 121
  Sipuncula, different types in, 164
  Sipuncula, staining reactions in, 167, 169
Acidophil granulocyte, 588
Acidophilic granular amoebocyte, 81, 82–83, 103, 249, 265
Acidophilic granule cell, 389
Acidophilic granulocyte, 164, 165, 167, 171, 245
Active phagocyte, 371
Adipohaemocyte, 335, 340, 341, 343, 344, 346, 374, 375, 382, 395, 426, 427, 434, 436, 439, 444, 479
  cytology, 342
  fat body cell, regarded as, 426, 436
  lipo-protein cell, compared with, 395
  modified granular cell, regarded as, 426, 436
Adipoleucocyte, 427
Adipo-spherular cells, 51
Adoptive transfer, 128–129
Agglutinin, 123, 290, 320, 321, 551, 609
  opsonic activity of, 220, 229, 409, 456, 633
Aggregation factor, Porifera, 24
Agranular leucocyte, 249, 255
Agranulocyte, 251
Alanine, 415
Alkaline phosphatase, 268, 275, 630, 631
  cellular and serum components, 293
  chloragosomes of *Lumbricus*, 106
  granulocytes of Sipuncula, 176
  peritoneal coelomocytes of polychaetes, 62
Allogeneic cell interaction, 614–617
Allogeneic graft, 28, 120, 123, 129, 133, 552, 553, 614
  non-fusion in colonial tunicates, 615–617
Amibocyte, 587, 589
Amibocyte à graisse, 589, 594
Amibocyte à vacuoles multiples, 589
Amibocyte de réserve à vacuoles, 589